Black Storm: Curse on the Caliphate

By: Robert L. Kelly

A PUBLICATION
JACK ASSBANKER™

Black Storm: Curse on the Caliphate

By: Robert L. Kelly

Library of Congress Control Number: 2014918745

Jack Assbanker™ Publications
Attention: Robert Kelly
1732 1st Avenue #24739
New York, New York 10128

A PUBLICATION
JACK ASSBANKER™

ACKNOWLEDGEMENTS

Thanks to all of my friends and family who have stood by my side, no matter what. It is only through this support and network of true love, was Colt Jackson ever possible to write and create. For those who read Colt's first mission, as a Commander in the U.S. Navy SEALs, you know he is a Christian warrior, battling evil. This action-packed thrill ride was written to immerse the reader directly into the action---all over the world. As a result, Blood Moon Over D'Apocalypse™ came to life and action hero Commander Colt Jackson, was born.

In this second book of the Colt Jackson Thriller & Adventure Stories, Commander Colt Jackson has been asked to do the impossible, once again. Black Storm: Curse on the Caliphate, is a teeth-rattling nail-biter. It realistically portrays the rigors and stresses placed on the men and women who fight for the freedom of people, all around the world. In this book, there is a global, sophisticated plan of attack against the United States by a highly organized army of Islamic fanatics. They are armed, dangerous, absolutely ruthless and wealthy---backed by the oil Sheikhs.

Curse on the Caliphate reflects the realities of what modern societies face today. With nearly an infinite number of threats posed against them, the use of futuristic technologies is vital to the national defense of all free people---and such technologies are ingeniously woven into the stories of Colt Jackson. Today's modern Jihadists have degrees from advanced educational institutions and have access to nuclear, biological and chemical weapons. Weapons of mass destruction can change, instantly, the way of life for millions of people. Only through the heroic actions of real men, as portrayed by Colt Jackson and his SEAL team, and real women, as portrayed by Ms. Lori Sanders and her associates at the NSA and CIA, is our great way of life possible.

I hope the actions and nature of the characters developed in the Colt Jackson Thriller & Adventure Stories, creates greater awareness and appreciation for all men and women in our military and intelligence services. They protect our great nation---and they all deserve a huge raise.

I thank God for putting Jesus in my life and giving me inspiration to create Colt Jackson and write The Federal Reserve Trilogy.

Robert L. Kelly
Author

About The Federal Reserve Trilogy:

Mr. Kelly's warnings relating to the approaching credit crisis in 2007 and 2008 helped some people avoid devastating losses in their portfolios, because they took action *before* the crisis erupted. As everyone knows, ultimately, there was a complete stock-market meltdown and difficulty! In a "Déjà vu, all-over again" moment, Mr. Kelly is sounding the alarm bells EVEN LOUDER, yet another time.

The Federal Reserve Trilogy issues an urgent warning to all people and documents the causes, the motivations and the outcome of a devastating economic firestorm, called "D'Apocalypse™." It is swiftly approaching and set to strike nations all over the world. The trilogy unveils the secret plans of the bankers and global elite to collapse the financial system and seize control of the world's monetary policy and money supply systems. Mr. Kelly sleuths out the facts, figures, history, charts and patterns, revealing their master strategy, while compellingly providing nations, corporations and all wealth classes, specific recommendations to avoid disaster during this upcoming grab for power. The end result of D'Apocalypse™ will be social devastation and a severe economic meltdown, purposely caused by the bankers and the elite. It will rain ruin on hundreds of millions of people–with the onset of a dangerous and real global war, the likely result!

There are very few people, or organizations, with the ability to analyze and distill sets of complex information into understandable, tactical and easy-to-implement action plans. Mr. Kelly amply demonstrates these abilities, with the writing of The Federal Reserve Trilogy.

The $30 Trillion Heist–Scene Of The Crime?, and The $30 Trillion Heist–Follow The Money!, spell out how the banks, the Federal Reserve and the elite heisted nearly $30 Trillion from U.S. taxpayers–without the knowledge, or consent of Congress. Importantly, the books uncover what they are doing with this heist money. D'Apocalypse™ Now! unveils the logical conclusion and aftermath of the great heist. Reading it will help families, hedge funds, companies, and nations, avoid the worst of this purposely engineered, financial catastrophe.

Fortunes will be made and lost during the Great D'Apocalypse™ and The Federal Reserve Trilogy empowers you to assess the facts and compelling evidence, allowing you to make powerful and intelligent decisions during the course of the next several years. It is a set of books you will want to have in digital format because of their robust links, marvelous images, graphics and artwork. You may also want a hard copy for backup, just in case the lights go out and things really get bad!

A Jack Assbanker™ Publication

TABLE OF CONTENTS

Black Storm: Curse on the Caliphate is a work of fiction. Names, characters, places, and incidents either are the product of the author's imagination, or are used fictitiously. Any resemblance to actual persons, living or dead, events, or locales is entirely coincidental.

Other works by the author, Robert L. Kelly, include Blood Moon Over D'Apocalypse™, a work of fiction and the first book in the series of "Colt Jackson Thriller & Adventures Stories." Mr. Kelly also authored The Federal Reserve Trilogy. The three books of this non-fiction trilogy comprise a breakthrough in investigative journalism on the subject of The Federal Reserve System. The first two books are titled, The $30 Trillion Heist---Scene Of The Crime? and The $30 Trillion Heist---Follow The Money! while the third book is called D'Apocalypse™ Now!---The Doomsday Cycle. The third book is a combination of non-fiction and prognostication about future events in the financial markets. It predicts a financial catastrophe will strike because of the Federal Reserve's actions during what the author calls "The $30 Trillion Heist." Everyone should be interested in reading about how the Fed, the big banks and the uber elite secretly heisted, without the consent of Congress, nearly $30 Trillion in taxpayer money---to bail themselves out from losses they incurred during the credit crisis.

Why We Need Heroes In Apaches

According to Islam, any religion which does not agree with it is to be conquered through violence. This is why Muhammad and his followers have converted nearly 1.6 billion people---23% of the world's population. Conquered peoples either convert, or die. The words, below, are directly from their Quran.

They are also why we need brave people like Colt Jackson to do battle with them.

Quran** 9:29 "29. Fight against those who (1) believe not in Allah, (2) nor in the Last Day, (3) nor forbid that which has been forbidden by Allah and His Messenger (4) and those who acknowledge not the religion of truth (i.e. Islam) among the people of the Scripture (Jews and Christians), until they pay the Jizyah with willing submission, and feel themselves subdued."

Islam is directly against Christianity because Christians believe in the trinity---three persons and one God. Quran 3:151 states: "Soon shall We cast terror into the hearts of the Unbelievers, for that they joined companions with Allah, for which He had sent no authority".

Quran 2:191-193 "191 And kill them wherever you find them, and turn them out from where they have turned you out. And Al-Fitnah (disbelief and worshipping of others along with Allah) is worse than killing…But if they attack you, then kill them. Such is the recompense of the disbelievers. 192 But if they cease, then Allah is Oft-Forgiving, Most Merciful. 193 And fight them until there is no more Fitnah and worship is for Allah (alone). But if they cease, let there be no transgression except against Az-Zalimun (the polytheists, and wrong-doers, etc.)."

* Apache photo is a work product of the U.S. Government and is a public domain image. Artistic affects by Robert L. Kelly.

** Translation is from the Noble Quran, by www.dar-us-salam.com.

CHAPTER 1

SEXUAL JIHAD

Modern day---July 16 and the last day of Ramadan, in the isolated Kurdish town of Rawanduz, Iraq, 167 kilometers, or 104 miles east of Mosul.

The lead cleric in the mosque with the white head piece repeatedly cries out "TAKBIR" and the crowd around him responds "ALLAHU AKBAR" until they are in a trance-like state and whipped into frenzy. "Allahu Akbar" has historically been a Muslim battle cry, during war. It was first used by the prophet Muhammad in the Battle of Badr, which was the first battle in Islam. The words mean "Allah is the greatest."

Standing beside the lead cleric, is the self-proclaimed Caliph, Abu-Bakr al-Baghdadi, a Sunni Muslim. Caliph Abu-Bakr is the Islamic State's leader. The Islamic State, based in the Levant area of the Middle East, includes Jordan, Israel, Palestine, Lebanon, Cyprus and an area in southern Turkey north of Aleppo. Abu-Bakr has declared the Levant Region a Caliphate, with religious power and authority over all Muslims, worldwide. He christened it the "Islamic State," or:

لامية الإس الدولـة, it is pronounced "ad-Dawlah al-'Islāmiyyah" in Arabic.

The Caliph is also wise to the world's ways. He holds a PhD from MIT, is a certified genius and is named after the Prophet Muhammad's advisor, and companion. He led Islam for the Sunnis after Muhammad's death in 632 A.D, or year 10 of the Islamic Hijrah lunar calendar.

Abu-Bakr is the man who first formed the Islamic State of Iraq and Syria, "ISIS" and has used his Mensa IQ to plan brutal murder, while following the writings in the Quran. As a result, he has created a formidable presence for himself in the Middle East and around the world.

Because of his many victories, he now calls his emerging nation, "ISIL," for the Islamic State of Iraq and the Levant.

The Caliph has also justified his claim of leadership, worldwide over

Muslims, by claiming he was given divine inspiration from an angel. The Caliph, in his powerful and booming voice, has told everyone the angel was sent by Allah Himself. The angel's orders for Abu-Bakr were to conquer the Middle East, conquer Rome and destroy the United States---the "Great Satan."

The great Caliph is constantly asked by his followers to recount the story of how Gabriel came down from heaven to speak with him. The great Caliph made the story his own and used it to his full advantage. The people worship him, even though his surge to power has come with some tension among other Muslim leaders in the region---particularly with Turkey and many of the oil-rich Islamic states, with the exception of Qatar and the U.A.E.

In the west, the Caliph is viewed as a radical and brutal Jihadist, with a great thirst for power. Among his fanatical followers in the Middle East, however, and among his followers around the world, he is considered the Prophet Muhammad incarnate.

Caliph Abu-Bakr looks over the crowd of adoring Muslims in the mosque. His scimitar is dangling by his waste and its curved, razor-sharp blade is protected inside of his sabre's sheath. The sword's mere presence is designed to intimidate and threaten anyone who might disobey him.

His dark, black stare moves the crowd and engages many people, directly in the eyes. The effect is a satanic stare---which is mesmerizing. The throng cries out even louder, at an even faster rate, "ALLAHU AKBAR! ALLAHU AKBAR! ALLAHU AKBAR!"

Then, his eyes swing down low, before him. Lying in front of his feet is a young girl, facing the crowd. She is fourteen years old and her head is bowed down on the prayer rug she has now made her own. As the crowd chants hysterically, her body heaves up and down. She is sobbing before the throng of worshippers. Streams of tears pour down her face and her nose runs with mucous and snot, uncontrollably.

The Islamic State has issued a fatwa ordering unmarried females to offer themselves as "comfort women" to the fighting men of ISIL---they are also to join the war, by entering into a sexual Jihad to pleasure the Caliph's army. Their Jihad and mission is simple. They are to boost the morale of the fighters in Syria and throughout the Levant, by servicing the militants sexually, in any way the men desire.

Upon hearing of the fatwa and the global, Islamic infatuation with this self-proclaimed successor to the Prophet Muhammad, women from all over the world willingly traveled to Syria and the Levant in the name of this sexual Jihad.

They give pleasure to fighters, on a daily basis throughout the region, with many of them each servicing over a dozen men a day.

Unfortunately, thousands of other women have been forced into the sexual Jihad against their will. As ISIL pushed forward and conquered cities, territories and regions, the conquerors' first task was the killing of all the men who did not join the Jihad.

A Man Being Executed By The Islamic State
Public domain image.

Upon entering a conquered territory, militants simply line up the captured men, who do not join the Jihad. Then, they summarily shoot them in the back of the head and bury them in mass graves. The Jihadists' second task is then capturing all the women in the city, or area, they conquer. When caught, most of these women are now "unmarried," because of the brutal murder of their men.

It turns out very few men are ever allowed to join ISIL from a conquered territory. This is because the Islamic State fighters' claim, over the conquered territory's women, highly motivates them to kill nearly every man they capture. The Jihadists want virgins in heaven, all right, but they also have a ravenous desire for sexual conquest on earth as well---with as many women as possible.

YouTube, CNN and the BBC regularly broadcast hundreds of hours of these brutal execution videos, but rarely do the reporters connect the dots as to why so many men are being murdered. The entire sexual Jihad among the ISIL warriors is a driving force which creates courage and motivation among the men. They simply crave sexual pleasure.

ISIL loves it when Western media broadcasts the footage of the executions, because it puts the fear of God---Allah---into the hearts and souls of every enemy they face. This has allowed the Caliph to create a legacy of capitulation by opposing forces who dare to face him. This makes winning battles and capturing unwilling women for sexual Jihad, a much easier task to accomplish.

In a diabolically clever PR scheme, ISIL deploys this psychological warfare against all its enemies at virtually no cost. Fighters merely take video of the tens of thousands of people they murder and simply upload the footage onto YouTube. It costs them nothing and creates hysteria and weakness wherever its army attacks. The weak Western media companies can't resist the gory and graphic video footage and they broadcast these slayings at will.

The Caliph knows the Prophet Muhammad would be proud of him.

The young girl with her back to the Caliph was captured here, east of Mosul, in the Kurdish region of Iraq, in what historically has been called, "Kurdistan." Rawanduz is nestled against bluffs, rolling hills and mountains in Northern Iraq. It is isolated and was easy prey for the ISIL army. Kurdish women are renowned for their beauty and the young woman in front of the Caliph is certainly no exception.

She is also a true virgin and was taken right out of her home, where, while she was being taken, she witnessed the execution of her father, who tried to protect her. The ISIL murderer warriors shot her dad in the back of the head. She will never forget him falling over like a lump of clay, with blood and brains sprayed all over their family's living room.

This girl is on the verge of becoming a full-grown woman. She caught the great Caliph's eye when he entered her home to take the spoils of war.

The problem was, when the Caliph was forcing himself on her, she scratched at him like a wildcat. The Caliph slapped her, but not hard. He wanted her to come to him of her own free will. He attempted, again, some days later to sexually assault her, but this time she bit him, drawing blood. She had been kept under guard, since then, because the Caliph wanted her all to himself. But now, the Caliph had had enough.

Earlier, as the men were gathering in the mosque, he brought her inside and forced her onto the prayer rug, facing the growing crowd.

He and the cleric had remained behind her. The Caliph was hoping she would change her mind and come to him after being exposed to the hysterical chanting and urging of the crowd of men. They urged her to join the Jihad and serve the fight, sexually.

To further try and motivate her, through shame, the Caliph wanted to show her what might happen if she didn't come to him, of her own free will.

Suddenly and lustfully, the Caliph rips off her burka. The Caliph grabs her with such force, her long, black-flowing robe flies through the air. The power of the Caliph's grasp on her clothing forces her to stand right up, revealing an exquisitely beautiful woman. She now stands, nearly stark naked, in front of three-hundred men. The men go crazy, continuing to strengthen their cries that God is great.

She looks like a princess, as many Kurdish women do. Her beautiful, but sad, dark eyes droop, like a young fawn, and her long black hair shines under the bright light of the mosque. She is just coming into womanhood and her breasts are clearly visible, bulging out of her brassiere, with the dark shape of her nipples clearly showing underneath the sheerness of the bra. She has a voluptuous figure and it is now very clear why the great Caliph wants her all for himself.

He truly desires her and his lust is obvious, as he leers over her nearly naked body.

She is begging for her life. "PLEASE, PLEASE let me go home to my family!" She then bows down to the Caliph and holds his feet. "PLEASE....P-L-E-A-S-E, have mercy on me. Let me just go back home. M-O-M-M-Y...D-A-D-D-Y...PLEASE HELP ME!"

The Caliph becomes aroused. He thinks, "*Should I show her mercy?*"

After more chanting and raging by the sexually starved and voracious crowd, the Caliph finally addresses everyone. It is now his turn to speak to the men and inspire them.

The young Kurdish princess continues to whimper before him.

The great Caliph proclaims with a powerful and confident voice, "As we stand, blessed by Allah on the 29th day of Ramadan, we are now standing in control of the entire Levant. The entirety of Iraq, Syria, Jordan and Palestine has succumbed to our attack---nearly the entire eastern Mediterranean!"

The mosque crowd immediately, screams, "Allahu Akbar! Allahu Akbar!"

Caliph Abu-Bakr continues, "Now, like Muhammad before us, when he began our march on the 17th day of Ramadan, 2AH* and divinely won the battle of Badr against the infidel Quraishi in Mecca, our revolution will push us even further! This time we will destroy the West! Muhammad was a strategic genius and like him, WE MUST ALSO BE!"

The crowd goes into a trance-like state upon hearing the powerful words of their great leader. The three-hundred men chant in a whirling dervish "Allahu Akbar! Allahu Akbar!"

Caliph Abu-Bakr declares, "I have prayed and have seen a vision from Muhammad, himself! I am selected to create an alliance with the enemies of the Great Satan! I was told by our great Prophet, 'The enemy of your enemy is your friend for this great battle! Afterwards, you will turn on them who do not believe in Allah!' Let us cry out to Allah at the end of this Ramadan, our most holy Takbir Eid al-Fitr!"

The crowd chants the Takbir of Eid:

> والله . الله إلا إلـه لا ،أكـبر الله ،أكـبر الله ،أكـبر الله"
> "الحمـد ولله ، أكـبر الله ، أكـبر
> Allahu akbar; Allahu akbar; la illaha il Allah, Allahu akbar; Allahu akbar, walilahil-hamd.
>
> (Allah is most great; Allah is most great; there is no god except Allah, Allah is most great, Allah is most great, and all praise be to Allah).
>
> الله ، الله إلا إلـه لا .. أكـبر الله .. أكـبر الله .. أكـبر الله"
> " الحمـد ولله .. أكـبر الله .. أكـبر الله .. أكـبر
> Allahu akbar; Allahu akbar; Allahu akbar; la illaha il Allah; Allahu Akbar; Allahu akbar; Allahu akbar walilahil-hamd.
>
> (Allah is most great; Allah is most great; Allah is most great; there is no god except Allah, Allah is most great, Allah is most great, and all praise be to Allah).

* "2AH" is the year according to the Islamic Hijri, lunar calendar. The 17th day of Ramadan, 2AH, is the equivalent of March 13, 624 AD, on the Western world's Julian Calendar.

الله ، الله إلا إلـه لا .. أكـبر الله .. أكـبر الله .. أكـبر الله “
“ الحمـد ولله .. أكـبر الله .. أكـبر
Allahu akbaru kabira; Allahu akbaru kabira; Allahu akbaru wa-ajall; Allahu akbaru, wa-lillah il-hamd.

> (Allah is the greatest - Great indeed; Allah is the greatest - Great indeed; Allah is the greatest and the Most Majestic; Allah is the greatest; and to Allah belongs all praise).

.الحمـد لله ملك ولله .أكـبر الله .أكـبر الله .أكـبر الله“
”لتوجيهنـــا أكـبر الله .المهيبـة ومعظم أكـبر الله
Allahu akbar; Allahu akbar; Allahu akbar; wa-lillah ilhamd, Allahu akbaru wa-ajall; Alahu akbaru 'ala ma hadana.

> (Allah is the greatest; Allah is the greatest; Allah is the greatest; and to Allah belongs all praise. Allah is the greatest and the Most Majestic; Allah is the Greatest for guiding us).

أكـبر الله ! أكـبر الله!
Allahu Akbar! Allahu Akbar!

> (Allah is the Greatest! Allah is the Greatest!)

Every eyeball is on the great Caliph. They want to see if the great Caliph will be strong and do what needs to be done.

Caliph Abu-Bakr turns the beautiful woman around and sees the magic of her beauty and the fullness of her ripe breasts. He is hoping she will ask for mercy and forgiveness, by joining him in the Jihad. He longs for her to come with him freely, of her own free will, to his tent.

He grabs her in front of the crowd and pulls her body close to his, forcefully, but with gentleness, hoping to have her feel his passion and desire for her. His coal-black stare penetrates her deeply, she feels weak inside. She trembles as he draws her lips close to his own.

The crowd looks at her curving figure and is salacious with desire. They are screaming at the top of their lungs for the great Caliph to take her before their very eyes and give them the leftovers.

He whispers to her, so only she can hear. "Will you come with me willingly, or do you want to be left with this pack of lecherous dogs?"

She allows herself to be taken closer to the Great Caliph. Her breasts rub against his strong chest. His pulse quickens and he is enamored with her beauty and the utter softness of her skin. He tells her, "You are to be mine! Let us go…"

Much to the Caliph's delight, the young maiden allows her breasts to push even harder against his chest, and her thighs rub against his rock hard penis. She says, while looking directly in his eyes and moving close to him, "Oh, great Caliph…"

The Caliph draws her lips close to his and starts to kiss her.

The strength of her bite and the blood which appears from his lip, runs out of her mouth, drips down her chin and drops down onto her breasts. The white bra is now splattered with the Caliph's crimson-red blood.

He violently throws her down onto the prayer rug and the crowd surges forward. Men in front are raising their robes, in anticipation of the great rape and violation which is sure to come.

The Caliph sees this and draws his scimitar. He holds it high over the virgin's head and swings down with vicious power onto the back of her neck. Her beautiful head is immediately severed from her body. It falls toward the crowd. The once beautiful face, of the lovely young virgin, stops directly before them.

Her fawn-like eyes remain wide open in death, staring at the crowd before her.

With blood dripping down from his bit lip, the Caliph raises his scimitar and rages,

"ALLAHU AKBAR! ALLAHU AKBAR!"

The crowd repeats the chant in a dizzying sight of swirls, dances and rabid chanting,

"ALLAHU AKBAR! ALLAHU AKBAR!"

CHAPTER 2

THE PEARL

One year later, on August 17th, an important meeting is taking place at the Marsa Malaz Kempinski Hotel at The Pearl. "The Pearl" is the globally recognized and famous, set of man-made islands off the coast of D'oha, Qatar.

The oil-rich nation of Qatar has spent billions of dollars creating this elaborate and miraculous engineering marvel. It is no coincidence the Marsa Malaz Hotel is its crown jewel. The Marsa Malaz is managed by the Kempinski Hotel organization, but is secretly owned by the ruler of Qatar, Emir Sheikh Muhammad Al Therani, who has helped organize the meeting.

The ultra-luxury hotel sits on its own private island, within the Pearl's complex of islands, and boasts beautiful Arabic design with Europe's finest architecture and engineering behind it. When the Marsa Malaz is approached from any angle, it looks like a palace, with its pink-pastel sandstone warming and beckoning visitors as they come near. It oozes elegance and majesty. The Marsa Malaz has 281 rooms, including 69 suites, 2 Presidential suites and 2 Royal suites. All of them feature sea or garden views, contain a lounge area, have spacious balconies, as well as world-class technology and butler services. The presidential and royal suites feature large private open-air terraces with pools, or jacuzzis.

The hotel's private beach area is a special one. Not only does it provide for a private entryway into the back of the hotel, but it has a spectacular outdoor swimming pool, constructed of silver-inlaid marble which was brought in from Italy. The hotel is also equipped with a special yacht jetty, tennis courts, water sports of all kinds, along with beautiful gardens and patios. Entertainment options include six restaurants, a lounge café, an "uber-upscale" bar, a roof-top nightclub, as well as a pool restaurant and spa, which is over 3,000 square meters.

Finally, the hotel has some of the Middle East's finest meeting facilities available, including a 1,100 square meter Grand Ballroom, famous for its sparkling white interior design setting, complemented with amazing

Baccarat Zenith Crystal Chandeliers, purchased at over $150,000 each. They were specially made for the Sheikh, in accordance with his demanding specifications. The chandeliers hang down from the ceiling like the Hanging Gardens of Babylon. They simply dazzle anyone who enters the room, which was clearly designed to impress. The Grand Ballroom boasts over $2,000,000 in lighting, making it one of the most spectacular places for a meeting, in the world.

This is the room, and this is the property, where a meeting of historic events is occurring. The Marsa Malaz is a palace and today it is hosting guests bent on the destruction of America.

Electronic draping and high security shields the meeting's attendees from any potential eaves droppers. In attendance are representatives from the governments and militaries of the new alliance, including The Islamic State, Turkey, Qatar, North Korea, Venezuela, Pakistan and splinter groups from China and Russia.

Despite the fact Abu-Bakr has proclaimed himself the head Caliph, Istanbul is the city which has been named the head of the Islamic caliphate, by all the Muslim scholars. This has presented some friction among the various factions of Islam, but everyone has unified under the goal of destroying "The Great Satan," The United States of America.

According to the cleric Sheikh Yusuf al-Masri, who is the Chairman of the International Union for Muslim Scholars, Turkey's leadership has been crucial in uniting all kinds of Muslims. Turkey's Islamic leadership's goal is simple. They want to rid the world of "The Great Satan." To accomplish this objective, the leadership has pushed an agenda calling for the unification of Wahhabist Sunnis, Persian Shiites, as well as the Ummah in Africa, Asia, Europe and other sects throughout the world. This has resulted in the building of an every powerful, Islamic State. The cleric Chairman proclaims, for the world to hear, the man who made this possible, and the man who changed Turkey, is Recep Tayyip Erdogan---Turkey's supreme leader.

Outside the Middle East, with Islamic fanatics all over the world utilizing tactics calling for brutal mass murder, genocide, rape, torture and mass executions, the world is waking up. Even Jesus Christ, speaking through the Book of Revelation, said Pergamum (modern day Turkey) is the seat of Satan:

> "[12]To the angel of the church in Pergamum write: These are the words of him who has the sharp, double-edged sword. [13] I know where you live--where Satan has his throne. Yet you remain true to my name. You did not renounce your faith in me, not even in the days of Antipas, my faithful witness, who was put to death in your city--where Satan lives" (Rev 2:12-13 New International Version).

In the Ballroom of the Marsa Malaz Hotel, a gathering of men is sitting around the elegant grand table. Caliph Abu-Bakr is first to speak. "Praise be to Allah! And welcome, all of you. The people around this table are chosen by Allah to rid ourselves of The Great Satan. Our handwritten communications and personal couriers have avoided the snooping ears of the NSA. Thanks be to Allah for the last two years, we have been completely protected. Today, we dare to meet in this electronically screened room---giving us bullet-proof security for this last meeting."

General Kim Se Bong, from North Korea, ruler of the Second Economic Committee, is the exporter of missiles and nuclear technology from North Korea. He responds to the Caliph and looks around the grand table, "Caliph Abu-Bakr, you have been wise in choosing your allies for this noble mission. For our part, we have completed the assemblage of the stolen enriched uranium from Iran and assembled the bombs and missiles to be launched from the ships."

Abdul Qadeer Khan then speaks, "As-Salmu Alaykum, Caliph Abu-Bakr. As the father of Pakistan's nuclear army, I have been pleased to teach some of the PhD's you sent to us. The Saudi students who trained at the American universities in physics and chemistry were particularly adept."

Caliph Abu-Bakr replies, "As-Salmu Alaykum, Abdul Khan. Your greatness precedes you. Thank you for teaching our followers the secrets of creating dirty weapons. When we capture facilities to enrich uranium and plutonium, they will be ready to assemble and build our own weapons. Insha'Allah. Vice Chief of Naval Staff, Admiral Muhammad Mubarak, we especially would like to thank the Emir and the military of Qatar. Your hospitality and help has been invaluable."

Admiral Muhammad Mubarak of the Qatari Navy speaks next, "As-Salmu Alaykum, Caliph. You know we are all on the Prophet's team. Elimination of the Great Satan will allow us to move west and conquer Greece, Rome, Spain and the rest of Europe. They will be overwhelmed--

-both internally by their foolhardy immigration policies and externally by our invasion. The ISIL army has electrified the entire Islamic State---and by this I mean every Sunni nation on earth!"

An olive-skinned man, in military uniform speaks next in a Spanish accent. He is General Jose Lopez, General-in-Chief, Operational Strategic Command of Venezuela. He states with confidence, "Venezuela stands ready with oil tankers for the missile launchers when the modifications are complete. These are expected soon and we can then fire them from the Gulf, as well as from the Pacific and Atlantic Oceans. Three of the missiles have been sent to us by General Bong and are steaming to Maracaibo as we speak. Sent in different directions, by three different ships, the EMP affect will be devastating to the natural gas, oil, defense and entire infrastructure across the south. The attack from the vessel from the Pacific Ocean, will be devastating to California. It will wipe out the electronics industry."

General Kim Se Bong states, "We have optimized the mobile missile launchers' capabilities by changing their angle of attack to maximize their thrust, using the most powerful solid-fuel boosters. As you all know, the higher we can detonate the nuclear devices, the broader and more severe the EMP impact will be. We are using the SA-2 launchers, which are mobile and easy to hide on a ship. We can also protect them from the prying eyes of NSA satellites and their ability to track radiation signatures."

Caliph Abu-Bakr replies, "You are to be praised to Allah. Truly your commitment against our common enemy, earns you stars in heaven, General Bong!"

General Kim Se Bong looks at the Caliph and says, "Allahu akbar, Caliph!"

Admiral Muhammad Mubarak then discusses his contribution. "As you all know, Qatar Chemical Company, QEMCO, does business all over the world. Our brother, and their CEO, Sheikh Sultan Bin Mohamed Rashid, has arranged for its distributor in Europe, LUBEROL, to deliver another set of missiles and payloads addressed to the Chemrol Corporation in New Jersey. Chemrol is their U.S. distributor and LUBEROL is regularly on the import/export records for U.S. Custom's purposes. LUBEROL regularly delivers high performance industrial lubricants, brake fluids, transmission fluids and all manner of advanced industrial oils. Of course, this shipment will be encased in lead and concrete, with a six-foot swimming pool wrapped around it."

Abdul Qadeer Khan is smiling, "Brilliant, my brother! You've made the fissile material double wrapped---making the Americans' detection systems useless. First the water barrier, then the lead/concrete barrier! Positively excellent."

General Jose Lopez excitedly says, "I can see the devious brilliance in this. When we send our ships to the Gulf, Charleston and the West Coast of America via the Panama Canal, we should use this same system. I assume you just drain the water and use the shipboard crane to take off the lead and concrete box around the launch pad?"

Admiral Muhammad Mubarak affirms the General's understanding, "Precisely. I have the plans with me, in case our great friends in Venezuela would like a copy of them."

"Yes, we would, Admiral! Eso seria excelente bien!" The General is very happy.

Sheikh Yusef al-Masri, Cleric Chairman of the International Union for Muslim Scholars, based in Istanbul, now speaks. "As-Salmu Alaykum, Caliph Abu-Bakr. The unification of the Muslim people, according to Allah the Merciful's will, is happening before our very eyes. The Islamic State and the delivery of justice for the Muslim world is the awakening. Our march through, and conquest over, the western European empire will commence immediately after the attack on America! By moderation, we encourage the unification of the Sunni and Shiite people into one powerful Islamic State. Allahu Akbar!"

Caliph Abu-Bakr responds resolutely, "Allahu Akbar! Allahu Akbar!"

Sheikh Yusef Al-Masri continues, "We have now attracted millions of Jihadi warriors to the Islamic State. We need them to take over Rome, Spain and all of Europe. The oil monies have allowed us to vastly improve our capabilities, and of course, we are blessed by Allah in the alliance with our Chinese brothers, particularly. There are over 50 million Muslims in China, alone!"

General Kim Se Bong responds, "And millions more throughout Asia, myself secretly included, of course! Allah the merciful may even grant us a chance to overthrow Kim Jong-un, the Supreme Leader of North Korea. He has no God and one day, he will meet his end. In the meantime, our influence grows and spreads throughout Southeast Asia."

Caliph Abu-Bakr looks at the General directly in the eyes and says, "Allah has blessed you with protection. Your influence on the military and Kim Jong-un has been spectacular. We are all truly grateful to you and Abdul Qadeer Khan for your help with the learning curve on the nuclear, biological and chemical weapons' front. They're our new sabers for the 21st century! Allah wills we show no mercy to those who don't follow our path. Allah particularly hates the Americans, who have spurned his righteous ways and turned to sin and perversion, while allowing the worship of many gods. The vilest, of which, is their worship of money. Allah abhors this."

"Thank you, my brother, Caliph Abu-Bakr. Whether I am from Pakistan, or you are from Saudi Arabia, it makes no difference," says Abdul Qadeer Khan, "We all are slaves to Allah's will and have a common enemy. This is the alliance which Allah provided us. We are all blessed at the dawning of our march! Imagine an army of 3 million Jihadists heading west! We shall sweep across the entirety of the soft belly of Europe in weeks! Insha'Allah."

Caliph Abu-Bakr shouts, "Insha'Allah!" (God willing).

Chinese Vice-Premier Wang Kai, is third-ranked in power behind Chinese Premier Zhang Li. He decides to now speak. "Gentlemen, on behalf of our allies and strategic partners in the great nation of China, we continue to commit ourselves to providing the brave Jihadists with the most modern of battlefield weapons. Our trading of technology and equipment for oil has served both of our interests. Our country is most grateful for the 10-year natural gas development contract you provided us to tap the massive reserves off of Gaza. Allahu Akbar!"

Sheikh Yusef Al-Masri responds, "Allahu Akbar, Vice-Premier! We are grateful to Allah for our Muslim brothers and sisters sprinkled through the ranks of the Politburo in China. Thank you for influencing the politics and policies of the great nation of China to assist our holy cause! Unifying the different branches of Islam, under one belief---albeit with some compromises, has made us all stronger."

Vice-Premier Wang Kai replies, "There is no thanks needed, my brother.

It is a calling from Allah. We are on the same team. Naturally, the Chinese are not ready to put their fingerprints on the EMP blasts and other non-conventional weapons the Jihadists will use. The Americans will still have formidable power after the attack---enough might in their submarine force, alone, to take out the entirety of China."

General Jose Lopez jumps on this point. "This, I can attest to. In my early years growing up in America, I was trained in their submarines. I can remember vividly their stealth capabilities and their nuclear arsenals on board. They are on alert at all times and never sleep. I believe in our Jihadist calling and felt it was my duty to return home to Venezuela, where I was born, to bring the word and work of the Prophet to this predominantly Catholic nation. We have tens of thousands of followers now!"

Caliph Abu-Bakr blesses him, "May Allah always bless the Prophet and his spiritual brothers!"

General Dmitry Zhirov is a Russian General who hides his Islamic, Chechen heritage from the Kremlin. "Allahu Akbar, gentlemen! I would like to say, on behalf of the great nation of Russia and our devotion to overthrowing the Christian Orthodoxy running our nation, Insha'Allah! The suitcase nuclear weapons obtained by Al Qaeda have been provided to us for this cause. Al Qaeda in Afghanistan and Pakistan hid them and sealed them from satellite detection in the caves. The tritium triggers had to be replaced and we had to replace a few of the mechanisms, but three weapons are ready for deployment. The CIA is already aware of these weapons, but has discounted their threat analysis because of the decaying fuses and other components vital to their detonation."

"Truly, Allahu akbaru wa-ajall---Allah is the greatest! This miracle you provide us General Zhirov, is a sabre of destructive power which shall rattle the American heartland to the core!" Caliph Abu-Bakr bows his head in respect to the Russian Muslim.

General Dmitry Zhirov then says, "Because the Americans know these weapons are in the hands of the Taliban and Al Qaeda in Afghanistan, they will blame them---and likely retaliate in this area. The thousands of our brothers and sisters who die because of a counter strike shall die martyr's deaths! Surely, they will have dozens of virgins in heaven for their pleasure, when they meet the Prophet."

Sheikh Yusef Al-Masri gets excited, "Allahu akbaru wa-ajall!! They will truly be lavished with the most beautiful of women, the most handsome of men and the greatest feast and celebration, imaginable."

General Dmitry Zhirov, "Like China, Russia CANNOT be directly part of any use of unconventional weapons. If Russia is fingered in this by the Islamic State, my brothers, I warn you---Mother Russia, with all of its might, will retaliate and turn on you like the nested and hiding viper it is. It will not risk a Nuclear, Biological, or Chemical attack on its people by a wounded United States. As a reminder, with Iran and the Shiites destined to be destroyed by a counter strike, Russia will lose enormous economic advantages it enjoys with Iran and Syria."

Caliph Abu-Bakr quickly responds to this great concern, "We have discussed this at length, General Zhirov---this includes me, the Supreme Counsel, as well as the Emir of this wonderful nation hosting us, today---Qatar."

General Dmitry Zhirov then asks, "What shall I report to the Kremlin?"

Caliph Abu-Bakr, with a big smile on his face answers, "Mother Russia is to be rewarded ten-fold for her loss---this is as Allah wills it and is what the Supreme Council decided. Mother Russia will receive all the nuclear contracts in the Islamic Caliphate and the Caliphate shall build an additional natural gas pipeline for Russian natural gas to our warm-water ports. Like Hong Kong, Mother Russia shall enjoy a 100-year lease on the right-of-way for the lands and the port facilities."

Admiral Muhammad Mubarak then adds, "Also, on behalf of the Emir of Qatar and every other country in the Islamic State, from now on, all of our military purchases shall be split between Mother Russia and the glorious nation of China! After the attack, 100% of Islamic State strategic weapons and materiel purchases shall be done with your countries!"

Vice-Premier Wang Kai is now also smiling, "Allah is great indeed! This is a magnanimous gesture by the Islamic State which shall be music to the ears of the Premier and the Politburo!"

"Likewise, Admiral Mubarak!" says General Dmitry Zhirov, "The wisdom and strategic shrewdness of the Supreme Counsel and its advisors is led by Allah's inspiration. You give us the icing on the cake to ensure the plan works! Insha'Allah!"

At 2:00PM, Qatar time, the Emir of Qatar, Sheikh Muhammad Al Therani, strides into the room, along with his body guards. Everyone rises to greet the Emir out of respect for his courage in supporting the Jihad.

Qatar has sent billions of dollars to the Islamic State and is the reason why the Emir's reign is secure. The security of his family's rule in Qatar is in direct contrast with the King of Saudi Arabia, Abdullah ibn Abdilazz. He has been overthrown as Saudi Arabia's head of state, absolute monarch, and head of the House of Saud. His beheading sent Monarchs fleeing in all directions.

Emir Sheikh Muhammad Al Therani looks around the room. He smiles broadly and warmly at the men, "As-Salaamu 'Alaykum - Peace Be On You! Welcome to our beautiful new hotel and addition to The Pearl here in Qatar! Truly you are the men who are doing Allah's and the Prophets most important work! I am merely a servant to you!"

Caliph Abu-Bakr, who is very close with the Emir, says, "Oh, Great Emir! We are humbled and honored by your presence. Without you, the Jihad would not have swept the Levant like wild fire! The Middle East and the Sunni way is the imprint for life going forward! Oh, great one, PLEASE bless us with your thinking."

Emir Sheikh Muhammad Al Therani, honored by the kind words states, "I only have a few words, as you all know my faith in your work is great. My words are directed to our great allies in the countries of Russia and China. I want to confirm for you, directly, what you have heard in this room. It is the bond of every nation in the Islamic State. I also bring two copies of the treaty and agreement for you to bring back to Mother Russia and the glorious nation of China. We realize your superiors do not have faith in Allah yet, and must rely on written, contractual agreements. Allah will continue to bless us and one day soon, we will not have to write anything down on paper, because men will always keep their word. This is what Allah has ordained for us. Insha'Allah!"

Vice-Premier Wang Kai is very pleased with this show of respect and delivery of the treaty and agreement. "As-Salaamu 'Alaykum - Peace Be On You, great one! It is only through Allah's hand, is it possible for one of the smallest countries on earth, to strike down the Great Satan! You shall surely go down in history with the great Prophet and a man of Allah for the ages. Your dedication and humbleness is truly sent directly from Gabriel himself."

General Dmitry Zhirov is also elated, "As-Salaamu 'Alaykum - Peace Be On You, oh mighty and great Emir! I agree with my comrade and brother, Vice-Premier Kai. We are all inspired by your leadership. You are truly the mouse who roared!"

General Jose Lopez, from Venezuela, says, "What an honor to meet you in person, great one!"

Sheikh Yusef Al-Masri states to the Emir, "Thanks be to Allah that Istanbul provides unification now, among all Muslims. Using a moderate approach at unification has helped our ranks grow into the tens of millions. The unification of the Sunnis and Shiites made this possible…but you Emir were the sparkplug which made this engine come to life. The dream you told us before 9/11, is coming true. You foresaw a great leader arriving to lead an Islamic State. By this proof, I submit you are a Prophet. Insha'Allah! Walilahil-hamd" (All praise be to Allah!).

Emir Sheikh Muhammad Al Therani claps his hands. "Now, I believe most of the details can be followed up by all of you in the next twenty-four hours. But now, you must allow me the opportunity to entertain all of you here at the new palace of a hotel, the Marsa Malaz. Last evening's dinner was to help us all get to really know each other better. I trust you enjoyed yourselves. We have the most beautiful women, the finest foods, music, entertainment and beverages of your choice. The Levant restaurant is already rated with Michelin three stars and serves the finest in Middle East cuisine. We have built, as a special message to the world and inspired directly from a dream I received, a Spanish restaurant, also receiving three stars. We must re-accustom our people to Spanish food before we retake the Iberian Peninsula! The Kingdom of Qatar also brings you brave men one final gift---ALL OF THE WOMEN YOU WILL ENCOUNTER IN YOUR STAY HERE ARE VIRGIN MUSLIMS. They have traveled here from all over the world in response to our fatwa to fight a sexual Jihad and comfort our warriors and leaders. Only the most beautiful have been selected for your pleasure. Insha'Allah!"

Everyone around the table smiles broadly and shouts, "ALLAHU AKBAR!"

At this precise moment, some of the most beautiful women on earth pour into the room. Caterers, musicians and chefs also flow into the great and beautiful Grand Ballroom of the Marsa Malaz Kempinski Hotel. The

Emir has spared no expense. It will be a celebration for the ages inside the secure ballroom.

Emir Sheikh Muhammad Al Therani continues, "When we finish the celebration this evening, you will all sleep and finish business tomorrow. Tomorrow afternoon, a fleet of yachts will secretly whisk you away, with umbrellas covering your exit, just as they covered your arrival. You have no need to worry about anyone seeing you while you are on The Pearl. Before your arrival on the hydrofoil yacht, the only road coming onto the island was closed and no one else is in the hotel. All rooms were blocked last year, in anticipation of this great day!"

Everyone replies, "Al-hamdu lillah! (All praise and thanks to God!)."

General Dmitry Zhirov looks around the room at the beautiful women and says, mostly to himself, "Truly, Allah is the greatest! Allahu akbaru wa-ajall!!" He is grinning from ear-to-ear.

General Jose Lopez just smiles at his Russian ally.

The men then all disperse among the women and receive each other with smiles and lots of congratulatory back-slapping.

CHAPTER 3

THE LIGHT OF PRISM

Three months earlier, on May 17, Lori Sanders, newly appointed head of the NSA's PRISM Internet Spy Operations at Ft. Meade, Maryland is busy at work. Ms. Sanders received this promotion to oversee all PRISM programs, after the assassination of her associate, Ben Moby, by assassins in the fall of 2015.

Her still beautiful, green eyes sparkle behind her ugly, horn-rimmed glasses. She wears the ugly glasses on purpose and always wears her hair up, by design. This hides her soft, long, sandy-brown, blond-streaked hair. On a beach, she has to beat back the men who flock to her, trying to vie for her attention. She also has quite a figure. Underneath her NSA top secret clearance and her meticulously kept, serious appearance, Ms. Lori Sanders is a true American bombshell. She is a real beauty.

Lori is also very smart. She dresses very conservatively to camouflage her slender, perfect and gorgeous physique. Lori prides herself on being only about business at the NSA---except, of course, when it comes to Commander Colt Jackson, a man whom she loves.

Unfortunately, another woman loves him, too, and fortunately Commander Jackson does not work for the NSA. This has allowed her focus at work to propel her to the upper echelons of the agency, at a very young age.

Lori is staring at her computer screen after one of her subordinates from Prism's Internet Tracking of Conventions and Hotels, nicknamed "PITCH," knocks on her door.

"Yes, what is it?"

"Lori, I found something curious," says Kristi Kruger.

"What do you mean?"

"Well, I was doing my monthly analysis for the elite hotels, you know the usual---PRISM was downloading the reservations and customer information from the global hotel reservation system directory and the major hotel chains, from their corporate mainframes. Then, something popped up in Qatar." Kristi raises her eyebrow when speaking to Lori.

Lori Sanders responds, "So, you did put the Michelin resorts in the Middle East on alert status? I had a feeling this would be a good idea, given the wave of violence and sweeping nature of the Jihadists' takeover of the Levant. No one here a year ago would dream the House of Saud would be taken down through a military coup. The gruesome beheadings of the ruling family were a warning to every other country in the Middle East. I will never forget King Abdullah screaming for mercy, just before his head was cut off on YouTube. Sickening."

"Your analysis, Lori, was also correct about ISIL---the Islamic State. You said it will turn on the royalty in the countries of the Middle East. From their prophetic viewpoint, they consider the ruling families and people who have taken the oil wealth for their own pleasure to be infidels and hypocrites."

Lori responds, "It's a phenomena happening all over the world---even here at home. Just look at what the bankers did to the western financial system. If President Diaz, Admiral Whistler and Colt Jackson hadn't stopped them, our economy would have been even more horribly plundered!"

Kristi Kruger is looking seriously at her boss, "I seem to recall you received the NSA's highest accommodation for your work on 'Operation End the Fed,' Lori. Weren't you right in the middle of the whole thing?"

"That's a story for a different day, Kristi. Suffice it to say, it was truly a team effort…" Lori winks at her. "…In any event, these Jihadists will pillage everything they can and force everyone to their will. They are first and foremost, Sunni Muslims, and are beyond hard-liners. If you do not believe what they believe, they are taught by the Quran, to do battle with you until you convert---or you are killed."

Kristi Kruger shudders and says, "They are raving fanatics."

"Yes, but like many enemies of the United States before them---whether it was the Japanese kamikaze pilots, German storm troopers, or even the British when we fought against them for our freedom---they *can* be defeated," Lori states with confidence.

"Can I speak off the record?" Kristi asks.

"Of course."

"I'm not as experienced as you, Lori. I know this, but I just don't understand how our country can defeat the kind of fanatic who sincerely thinks if they kill one, or more of us, they are going to heaven---with dozens of virgins to boot! I mean admit it, we're completed disjointed on many levels---special interest groups tear us apart."

"Kristi, you have put your fingers on the pulse beat of the problem. Until America and every non-Muslim nation wakes up and realizes this is a religious war, fought by zealots who have been tricked into thinking they are fighting for God---but are truly fighting for Satan, we will lose."

"WE WILL?"

"Yes, Kristi. We lose ground, we will lose allies and we will lose in nearly every country on earth where they focus their fatwas to conquer territories they believe Allah wants conquered. It is a very powerful psychological weapon. Muhammad invented it on the 13th day of March 624 AD, which is the 17th day of Ramadan, in the year 2 AH."

"What does AH mean?"

"It is an abbreviation for Anno Hegirae, Latin for "Year of the Hijrah." The Muslim calendar dates from the Hijrah, 622 A.D. This is when Muhammad fled persecution from Mecca to a safe haven in Medina."

Kristi asks, "What happened on the 13th of March in 624 AD?"

"Once you understand this, you will understand why the Islamic State and nearly all forms of Islam, when you dig deep enough, espouse and APPROVE of violence to overthrow and destroy those who do not ascribe to their beliefs."

Kristi tells Lori, "You know, I did a paper on modern Islamic violence at Yale. Did you know the list of Islamic bombers during the last twenty years is downright scary? The Shoe Bomber, Beltway Snipers, Fort Hood, the Underwear Bomber, U.S.S. Cole, Madrid Train, Bali Nightclub,

London Subway, Moscow Theatre, Boston Marathon, Pan-Am Flight #93, Entebbe, Iranian Embassy takeover, Beirut U.S. Embassy bombing, Libyan Embassy attack, Buenos Aires, Israeli Olympics, Kenyan U.S. Embassy, Saudi Khobar Towers, Beirut Marines, Beslan Russian School, First World Trade Center Attack, Bombay & Mumbai India Attack, Achille Lauro Cruise Ship, as well as 9/11!!"

"I knew I hired you for a reason! You have one of the best memories we have ever tested."

"Thank you, Lori. I think these facts just stuck in my brain. They are compelling and overwhelming evidence of a global attack on western civilization by Islam."

"To answer your question, on the 13th day of March in 624 AD, which was during the Muslim's holy month of Ramadan, The Battle of Badr was fought between Muhammad and the Quraishi. The Quraishi were powerful merchants who controlled Mecca and the area surrounding it. Mecca was one of the wealthiest cities in the Middle East and its residents wanted nothing to do with Muhammad, or his beliefs."

Kristi Kruger asks, "Was it an even match?"

Lori continues, "Muhammad was outnumbered, according to Hadith tradition, by fifty-to-one in horse power and three-to-one in manpower. The Quraishi's created a strong defensive line, but Muhammad's Jihadists focused on a single point and attacked. They killed many important Quraishi leaders. This included Abu Jahl, who provided the main opposition inspiration against the Muslims. Muhammad learned then and there the power of fear and death to control people. With a small army, he took control of one of the world's most wealthy population centers and took 20% of the spoils for his booty."

Kristi exclaims with some surprise, "That's double the Biblical tithe of 10%!"

"That's right, but this entire lesson was written in the Quran. It states, "O Prophet, urge the believers to battle. If there are among you twenty who are steadfast, they will overcome two hundred. And if there are among you one hundred who are steadfast, they will overcome a thousand of those who have disbelieved because they are a people who do not understand." That's a direct quote from Surat Al-'Anfal 8:65. Surat Al-

'Anfal means "The Spoils of War."

"I see. They use this same battle cry to justify the sacking and taking of oil, money, gold and plundering of wealth and property from those they conquer. Genghis Khan used similar tactics to conquer half the earth!," Kristi says.

"Until America gets really hammered at home by them---even worse than 9/11, nothing much will change. Our country just doesn't come together on issues outside our borders until we are overtly threatened and our freedoms are taken away."

"I understand, Lori. Thanks for taking the time to explain this to me. We are sitting on an absolute powder keg in the Middle East---the more I know and understand, the more frightening the situation becomes. It sure gets me motivated, though, to keep tracking these bad guys!"

"EXACTLY! You're doing excellent work, Kristi. Two years out of Yale---you are doing exceptional work. Now, show me what you have found at the Marsa Malaz on The Pearl."

Kristi Kruger is looking stunned. "Just how did you know this was the hotel which popped up on my surveillance flag?"

Lori Sanders, a real studier and thinker, replies flatly, as if everyone should know this, "Because it is secretly owned by the Emir of Qatar, Sheikh Muhammad Al Therani. It is the newest jewel of the Michelin Hotels AND the Sheikh is up to his eyeballs in financing radical Sunni Jihadists all over the world. It seems like the logical place where something odd would happen. It is also completely surrounded by water and protected. There is only one road in and out. It is built as a Kempinski, for 'cover,' but through a massive contract provided to this hotel group, The Sheikh was secretly given the keys to this hotel. It is a Kempinski in name only."

Kristi is stunned by Lori's brilliance and mind. "I am SO happy you hired me. I have a lot to learn!"

Lori responds, "You're doing great, Kristi. Now, tell me what's going on..."

"Well, last month, the reservations in the Middle East hotels were basically empty in July---it is scorching hot there in the summer and everyone heads to London to go to Harrod's." Lori shoots her a stern look. "Ok, that's a bit of an overstatement, sorry. Historically, though, June through August is very, very slow for these hotels."

"Ok, I'm with you so far…"

Kristi continues, "This month, the Marsa Malaz, which is the jewel of The Pearl, shows it is *fully booked* for the period August 16 through the August 19."

"Have you cross-checked the convention logs? Perhaps the chamber of commerce for D'oha managed to create an event."

Kristi is way ahead of Lori on this, "I checked the logs and there are no events planned for the dead heat of August in D'oha, Qatar."

Lori then presses her questioning of her inexperienced employee, "What about the "Traceback.exe" program? Did you run it?"

Kristi nods her head. "I did. This is where it gets interesting. Usually, when a hotel gets fully booked, there is an enormous amount of information which is downloaded to the NSA storage systems in Bluffdale, Utah."

Lori Sanders says, "And?"

Kristi replies, "No such records exist!"

"What do you mean? Are ghosts staying at this hotel?"

"No, Lori, I mean there is just one single entry, which originated from a computer screen in the Emir's office! It blocked out all the rooms and even kicked out about one hundred pre-registered guests. I tracked the phone records from the Marsa Malaz Kempinski reservation system. Sure enough, phone calls AND emails were sent out to these pre-registered guests indicating their stay had been cancelled due to a needed upgrade and repair of a building wall. Each guest was invited to stay, any weekend, for the next year---with all expenses paid, except air fare! Highly unusual, don't you think?"

Lori says, "Especially for a brand new hotel with a three-star Michelin rating. We need eyes and ears at this hotel during this period. Please do me a favor and call Commander Colt Jackson in for a top-secret meeting.

He is a good out-of-the-box thinker and has healed up from the attack on the Federal Reserve and those mercenaries last year. His SEAL team and Intel unit were assigned the task of tracking down radical Muslim cells in the U.S. and western Europe."

"I will, Lori. Immediately."

Kristi Kruger leaves the office. Lori looks somewhat wistfully in the distance.

One day later, on May 18, after the request for an urgent meeting, Colt Jackson arrives at NSA Headquarters. He is a U.S. Navy SEAL Commander and Colonel in the United States Marine Corps. He is also a Medal of Honor recipient for his heroic actions during operation "End the Fed."

Colt and his team single-handedly prevented an economic disaster in America, which was planned by the elite banking families. His actions directly prevented "D'Apocalypse," which was planned to be a great debt and derivatives collapse by some of the wealthiest and powerful people on the planet. Colt stands physically strong, like a decathlon champion, is an avid patriot, and just like Lori Sanders, is a committed Christian and loves his country. When he greets people, he greets them with a steady gaze and his cobalt-blue, steel eyes never waiver. He has a charming smile and has brown hair. He is also very gregarious, with a sense of humor.

He is also a torn man in his personal life. He cares for two women and just can't seem to diminish his feelings for one, in favor of the other. As a result, Colt has buried himself in the work of the country and in healing himself from the horrific wounds he suffered during the attack on the Fed. Thanks to a dedicated workout regime, he is back 150%. He is in better shape than he ever has been in his entire life.

A strong feeling awakens inside of Colt as he approaches the Glass House in Ft. Meade. Lori Sanders needs him---Colt knows something is wrong. He knows she cares for him, but Lori is all business at the NSA. He knows she wouldn't call a meeting just to meet and greet. As Colt approaches the building, deep in thought about Lori, he just shakes his head. He thinks, speaking to himself before entering security at the NSA "*Ahhh, Lori Sanders...here we go again. I've sure missed you.*"

Inside NSA Headquarters, Colt's tall, athletic frame glides through the "Glass House." He is escorted to a secure conference room.

The room is ten stories below ground level, right near where the advanced supercomputers are located. Colt remembers seeing the beautiful Cray Computers and other advanced supercomputers when he was shown around, once before. The supercomputers are held in a huge room. But, Colt hears these floors at Ft. Meade, are only about one-tenth the size of the each new floor at the data center in Utah, which is truly massive.

As Colt enters the room, he stops dead in his tracks the moment he sees Lori Sanders. She is sitting, awaiting his arrival. Her disguised beauty is breathtaking to Colt. He knows her smell, he knows her kiss, he knows her smile and he knows she cares for him. He cares for her, too.

Colt Jackson is standing. Suddenly with weak knees, his cobalt-blue eyes are riveted on Lori. He smiles broadly. Colt doesn't even realize there is another person in the room.

"Hello, Lori. I have missed you!"

Lori Sanders replies, "Hello, Colt. I've missed you, too. You look terrific. You're not even limping anymore!!"

"I've actually improved my '40 time to 4.6 seconds---pretty good for a guy my age!"

Lori Sanders, "Oh, Colt, I know better! Thirty years of age is hardly old! Let me introduce you to my colleague, Kristi Kruger."

Colt immediately snaps back to reality and studies Kristi. Colt learned how to read people a long time ago. He sees yet another beautiful, fresh face in the room, albeit she looks right out of school. Lori had told him she hired some extremely bright people, from Harvard, or Yale, he couldn't remember which. He thinks, "*This must be one of them.*"

"Hello, Ms. Kruger, pleased to meet you." Colt extends his hand to shake hers.

"Please call me Kristi, Commander."

"You can call me Colt, everyone else does."

"Thank you, I have heard about you and what you did for the country last fall. Thank you. The plan the President came up with for the Greenbacks is working and the stock market is roaring again. The bankers are being jailed and the poor people of Detroit and other cities which were living off food stamps---can actually find work, again!"

Lori Sanders turns her attention to Kristi, "Kristi, you should know those were Colt's ideas. A lot of the credit goes to everyone on the team, of course, but Colt thought of the ideas to emulate Abraham Lincoln and Julius Caesar. President Diaz created the Greenbacks, which no longer pay interest to bankers, and he simply copied what Julius Caesar did two thousand years ago."

Kristi asks, "What was that?"

Lori answers, "Caesar ran into an economic depression, also. The mortgage holders wanted to kick everyone out of their homes. Caesar didn't allow this, but allowed the people living in the homes to write their own mortgages and allowed their past interest expense payments be CREDITED to their mortgage, reducing immediately the amount of principal owed."

Kristi is thinking, "You mean I have Julius Caesar to thank for the reduction on the principal my parents received on their home in Las Vegas?"

"Yes, and Colt Jackson, too!"

Kristi is somewhat shocked at the revelation Colt came up with these brilliant plans, she is growing infatuated with Colt and says, "That single-handedly helped 40 million people who were tossed out on the streets, Colt. It also helped every home owner achieve what was most sorely needed to restart the economy---debt relief! I heard you were like an action hero, but a boy genius, as well?" She smiles widely and mischievously at Colt.

Lori Sanders gives Kristi a look which could kill. Lori feels completely jealous, but her professionalism puts it aside, for now. At the right moment she will tell Ms. Kruger to back off, in her own way. Lori can be very persuasive, when she needs to be.

"He is, Kristi. This is why we roped him into this meeting. He has also been put in charge of tracking radical Islamic cells in the U.S. and Western Europe, while he was in recovery. The timing on this couldn't have been any better. He is the best man for the job."

Kristi says, "I get it. Shall I go through the briefing now?"

"Yes, please. Start from the top and tell him about the Emir and the strange activities we've uncovered at the Marsa Malaz in the middle of The Pearl in Doah, Qatar."

"OK, this is what we know..." Kristi goes through the prepared Power Point briefing. "As you can see, there are no corresponding arrivals for people on airplanes, or ships, during this period."

The briefing includes photos and imagery of the Kempinski Marsa Malaz Hotel, satellite photography of the single road onto the island, copies of the contract between Emir Sheikh Muhammad Al Therani and the Kempinski organization, as well as the history on reservations, and the fact the Emir, himself, blocked the rooms. Kristi says the obvious, but important, "August is not convention time in D'oha. It's just too hot."

Lori continues, "Given Sheikh Muhammad's aggressive activity in funding the Islamic State and Jihadists around the globe, he is someone we believe should be watched carefully."

Colt interjects, "I couldn't agree with you more, Lori. Thank you for the briefing, Kristi. You did an outstanding job and did very good intelligence work."

Kristi is beaming from head to toe and is obviously smitten by the handsome, Colt Jackson.

Lori says, "Yes, you did a good job, Kristi. You can go back to your desk now. I will see Commander Jackson out."

"Thank you, Lori. Thank you, Colt. I look forward to meeting you, again."

"Same here Kristi. You kids out of school are getting way too smart!"

Kristi tries to act very mature, not wanting to seem too young for the handsome man before her. "I've been here a couple of years now, so, I'm getting there. All of us were twenty-five sometime, right?" She looks at both Colt and Lori, who are each just over 30.

Colt answers, "Oh…You're older than I thought." He smiles at her and she smiles back. Lori then interrupts.

"Yes, indeed! Great job, Kristi. I'll speak to you later." Kristi departs the meeting room.

Colt Jackson turns to Lori Sanders and says, "Interesting briefing."

"She's taken with you, Colt! She was nearly giddy at the end!"

"Ahhh, Lori, she's just a kid and besides, I'm not interested. I have other things on my mind." He stares unswervingly into Lori's bright, emerald-green eyes.

"Cut it out, Colt! Not here. I can't handle it. Let's talk about this situation at the Marsa Malaz. I have a bad feeling inside."

Colt becomes serious, "I hate it when that happens, because I know you are usually right on target. Let me tell YOU, not her, what I know."

"OK, shoot."

Colt explains, "We have a mole. He was recruited when he was in the Navy. He even went to Annapolis and rode in nuclear subs for a few years. He is also a Muslim."

Lori is in a bit of disbelief. "You've got to be kidding me. WOW…"

Colt replies, "No, I'm not. He's an American citizen, but was born in Venezuela. He was taught Arabic as a child, as well as Spanish and English, of course. The CIA recruited him when he was in the Navy. He is one of the rare breeds of Muslims who doesn't believe violence is justified in order to convince people to convert to Islam."

"Where is he now?" Lori wants to know.

"About ten years ago, Admiral Steven Whistler---you remember my dear friend, don't you?"

"Of course Colt. You should know he was the one who backed you 1000% when nearly EVERYONE thought you had run from the battle at the Fed, when you snuck behind the mercs and then took them out."

"Well, that doesn't surprise me. We've certainly chewed some of the same ground before. Anyway, the Admiral, with his foot in both Naval Intelligence and the CIA, came upon this man's file. His name is Jose Lopez. The Admiral thought he could be a brilliant double agent---IN VENEZUELA. With his unique military training, Annapolis education and his Venezuelan citizenship, he was a natural."

"Wow, your Admiral is a smart cookie."

"He didn't get to be Admiral at his young age for nothing. Under the Hugo Chavez regime, Jose Lopez returned home and simply enlisted in the armed forces. His file quickly caught the eye of superiors and he was given the responsibility for building up the Venezuelan Navy."

"No match for ours, no doubt," Lori laughs.

"No…but it put him in the middle of an intelligence gold mine which is still paying dividends. When the new President, Nicolas Maduro, got into trouble economically through his radical socialist policies, Jose Lopez supported President Maduro---unwaveringly. In fact, it was Jose who fingered three generals, Oswaldo Hernández Sánchez, José Machillanda Díaz and Carlos Millán Yaguaracuto, as well as thirty others, for arrest by the President. Maduro is paranoid and doesn't even realize he arrested the wrong guys."

"Once more, WOW!"

"Part of my work the last six months is coming to grip with my assets available for detecting these radical cells. My work is taking me into covert intelligence and my entire SEAL Team is being trained in these techniques, accordingly. Everyone spent the last three months at Langley and on the farm in Virginia. It has been an eye-opener. Working and training at the same time, while rehabbing, has kept me beyond busy.

"I see," Lori says.

"This is also why I haven't seen you for a while, it's not because of Mary Barrister, or shall I say, not because I'm seeing Mary."

Mary Barrister also cares for Colt and Colt has feelings for this woman, as well. They lived through a harrowing crisis during the build up to Operation "End the Fed," last year. Mary's life was threatened multiple times, as was Colt's, and Mary lost her husband. He was one of Colt's dear friends. The information she delivered to Colt helped save the country from certain economic doom by the bankers and greedy elite behind the scenes.

Lori is unable to stifle a big smile at the news Colt is not seeing Mary Barrister. She becomes warm and sympathetic, "You poor thing, they've been running you ragged! I thought MOH winners wrote their own ticket?"

"My ticket is already written for me, Lori. I'll serve my country, and our Lord, to the end of time."

"I know. And I know you, Colt Jackson. I'm so proud of you." Lori cannot help herself.

They are still in the enclosed secure conference room and she runs into Colt's arms. She kisses him compassionately.

Colt holds Lori firmly against his body. He feels the warmth of her breasts rub against him. He feels a surging tide rush inside his body, and knows he could go completely out of control. After a long, compassionate kiss, they both gently and reluctantly pull away from each other. Colt is at a loss for words and filled with fever-pitch emotion. He stares deeply into Lori's eyes. Lori melts inside.

I hate to copy Maverick, but in his famous words, "Yo! Bartender!! Give me a beer to put these flames out!" Both Colt and Lori laugh.

"You're not exactly the romantic type, are you Colt Jackson?"

"Ahhh, Lori, you know me all too well. But, maybe someday I will surprise you. First, let me tell you the rest of what I know. OK?"

Lori feels every hair on her body rise and feels her breasts are still welled up. She loves this man, so much! "That's a good idea. I'm sorry I attacked you, Colt, I just couldn't help myself."

"Lori, it's a two-way street! I love being close to you and I love working with you. Things just seem to click when we're together. I truly love being able to care about you and also work with you---without one interfering with the other."

"I know, Colt, I know."

"Now," he says with a deep inhale, "back to Jose Lopez. It turns out, Jose is a natural. Admiral Whistler saw this from a mile away when he recruited him. Over the last several years, with Jose's Muslim background, he worked his way into the upper echelon of ISSA---the Islamic State of South America. It is in its early formation, but word got back to the Sheikh in Qatar and other key members of the Islamic State."

"You've got to be kidding me!"

"I kid you not. I met Jose in Aruba, just off the coast of Venezuela, last month. He is a great guy and we hit it off incredibly well. He told me something big was going down with the Islamic State. It was transforming itself---combining Turkey, ISIL, Saudi Arabia, Qatar, the U.A.E., Venezuela, Pakistan and North Korea. Now get this---the group also includes key members of the military and political parties from BOTH Russia and China!"

"We haven't picked up on ANYTHING of this at the NSA! How are they communicating?"

"After Snowden, everyone went to ground. They only communicate by hand-written communication and in-person meetings. The meetings are always protected by intensive electronic countermeasures, including jamming, copper screening, etc. The revelations of the NSA's capabilities changed communication behavior, worldwide. I'm afraid it's at least partially back to the old days---good old-fashioned, boots-on-the-ground spying."

"This is really important, Colt. Can I speak with General Maxwell about it?"

General Shane Maxwell is head of the NSA and worked directly with Lori and Colt during operation "End the Fed".

Colt replies, "Absolutely. Eyes and ears only, though. Nothing is to be put down in writing. OK? This resource, Jose Lopez, is now General-in-

Chief of Operational Strategic Command for the Venezuelan Armed Forces. Nothing can be left to chance with the protection of his identity as a U.S. intelligence agent and his commitment as a Muslim-American to freedom and peace."

"You know General Maxwell. He won't breathe a word, but this will help us tremendously in focusing additional resources on D'oha and The Pearl during the next three months, before anything goes down there.

"Perfect, Lori. I will brief Admiral Whistler and will be ready to make a move on this August date you discovered for the Marsa Malaz at The Pearl, in D'oha. We need to know immediately what is going down there, if anything. We better leave the conference room now. Eyebrows will start rising if we're alone too long in this room."

Lori Sanders sighs wistfully. "Always throwing cold water on the party, aren't you, Colt?" She smiles her dazzling smile and her eyes flash nothing but raw, pure emotion at Colt.

"Ah, Lori..." Colt laughs, pauses and then asks, "…would you go to dinner with me at the end of the week? Maybe on Friday, or Saturday?"

"I thought you'd never ask! Let's go for Saturday. This week is building into a frenzy, I have a feeling."

"Perfect. Pick you up at 7PM?"

"That would be great." Lori flashes her dazzling smile. They leave the room and walk to the elevator.

Colt exits the building and drives away from the Glass House. He looks at the building in his rear view mirror, as he makes the turn onto Canine Drive and heads to the Baltimore-Washington Parkway.

CHAPTER 4

A SURVEILLANCE PLAN

At the Pentagon in Admiral Steven Whistler's office, it is Wednesday, June 15, one month after the meeting between Lori Sanders, Colt Jackson and Kristi Kruger at the NSA.

In the office are Colt Jackson, Lori Sanders and General Shane Maxwell. They are discussing plans to monitor the comings and goings of the upcoming meeting taking place at the Marsa Malaz Hotel at The Pearl, in D'oha, Qatar.

Admiral Steven Whistler concludes his point on an issue they are discussing, "It's settled then. Colt, you'll take SEAL Team 3 out of Coronado, California, to Bahrain. With SEAL Team America and SEAL Team 3 jointly training in covert intelligence at Langley AND the Farm the last three months, SEAL Team 3, led by you, is a perfect fit for this mission.

"Copy that, sir. We worked well, together."

The Admiral continues, "Bahrain is only about 60 miles, as the crow flies, from D'oha. You will deploy via submarine and launch from the USS Minnesota in the new Shallow Water Combat Submersible (SWCS). They are light years better than the old SEAL Delivery Vehicles. The water is very clear and very blue around The Pearl. It is also extremely shallow. You will find this vehicle will help you navigate these challenges."

"Roger that, sir. Why is it so shallow?"

Lori Sanders answers the question. "The Pearl is actually built on a sandbar where low tide, naturally, would only be about one-half meter. When they constructed it, they had to build dykes and drain the area with hundreds of pieces of earth-moving equipment. This allowed them to make it a bit deeper for recreational water sports around the island, but the natural depth of the Gulf in this area is still extremely shallow.

Colt replies, "I remember there was a huge construction project to build

The Pearl, but I had no idea it was that shallow."

"It was actually the site of a pearl bank. In fact in years gone by, Qatari free divers were renowned for their ability to hold their breath," Lori replies.

General Shane Maxwell has something of a gift for Colt Jackson and his SEALs. He begins to speak, "Admiral…of course, this is your call, but I'd like to provide Commander Jackson's team with a new technology we developed in conjunction with the Defense Advanced Research Project Agency…DARPA. It's a Robo-vision system the NSA deployed for all-weather communications operations. It's waterproof."

The Admiral smiles and is pleased with this interesting contribution to the mission. "Please, by all means, General!"

"Perfect. Our R&D team adapted the Q-Warrior headgear developed by BAE* systems---for water use. It is exceptionally powerful. It allows personnel tracking, hostile enemy detection and incredible night-vision, with HD color feeds. Not only can you broadcast information to us, but we can send you information---yes, this means high-definition aerial images from drones and satellites."

Colt says with a bit of laughter, "This will be a huge upgrade from our old gear."

Lori starts detailing existing surveillance in place by the NSA and CIA at the Marsa Malaz. "There is only one road onto the island of the Marsa Malaz. We have it completely covered by surveillance via drone and hidden cameras. When you drive up to the entrance of the long entry road, you are welcomed by a "MARSA MALAZ" sign etched into the facade of the beige entry wall. We have hidden cameras embedded there, as well as in the tall street light posts which line the street.

Entrance Marsa Malaz
Doha, Qatar-The Pearl
Public domain image.

The street lights were perfect---the lights are held together by ornate black

* British Aerospace Systems

metal. Our cameras actually fit just beneath the glass, so the light is behind the camera eye. It provides a near perfect angle to capture the license plates of passing vehicles. These were placed there during construction and have not been activated, but have been tested. They work perfectly."

"Now, that's impressive," Colt answers as he smiles at Lori.

"After reviewing all the options, we came to the somewhat obvious conclusion the meeting will not have its attendees delivered by car, plane, or chopper---they would rouse too much suspicion. To avoid detection, we believe the attendees will arrive inconspicuously by boat. The hotel is on an island and sits among thousands of the wealthiest people in the world."

Colt follows the line of reasoning and says, "So, the most inconspicuous way for them to arrive would be by...yacht??"

"Exactly, Commander," General Maxwell chimes in. "If this is truly an event hosted by the Emir, they will arrive by ROYAL YACHT, or some kind of very expensive yacht which blends in with all the rest of the amazing craft in D'oha."

"Makes complete sense, General. Admiral, I have just a few questions."

"Yes, Colt?"

"Sir, I can see on this mission we are going to be spies. Do we have any orders to ENTER the hotel and attempt to plant bugs? Even if they have their meeting room screened off electronically, we could bug the entire facility and perhaps catch snatches of conversation which might prove valuable."

The Admiral turns to the General and asks, "General, what do you think, is it worth the risk of exposure?"

"With your man on the inside, Jose Lopez, I don't think it is. I think we might be better served to truly be covert spies on this one."

"I agree," says Lori, who is secretly relieved Colt Jackson will not be in danger.

Colt then asks perhaps, another important question. "Admiral, what are our rules of engagement? If this meeting is as important as we think it is, they might have the entire approach from the Persian Gulf patrolled."

"It's pretty simple, Commander. You are not to fire, unless fired upon."

Lori Sanders adds, "Admiral, if I could add just one more bit of Intel for Commander Jackson. The Marsa Malaz is connected to the Venetian District of the Pearl---it's called the Qanat Quartier, to be precise. There are thousands of eyeballs which will be looking down from the pastel-colored apartments and townhomes ringing the water---they literally gaze upon the hotel. Many people will also be out on the walkways, as the evening cools off. They crisscross the area and will truly remind you of Venice. You'll have to be weary of these random people, as well as the many boats passing by."

Colt thanks Lori, "Thank you, Lori. There is nothing worse than a thousand set of eyeballs when you're trying to spy!"

The Admiral speaks with authority and concern to Colt, "All the more reason to play it safe, Colt. We don't want people dialing the Qatari equivalent of "911" if they see frogmen climbing up the boulder sidewalls of the Marsa Malaz fortress. Your best bet will be to deploy your team in two areas. The first one should be along the beach entrance. It does not face the Venetian district. The second is the dock area. It is not actually attached to the island of the Marsa Malaz---it is part of the Venetian district. If the leaders arrive there, they have to take a vehicle up the lone entrance to the hotel. Unfortunately, on the other side of the channel is Fanar Fayrouz. Multi-million villas lie on that stretch of land---with a continuous and seamless set of concrete stairs which drop down into the cove."

Lori speaks seriously to Colt Jackson, "As you can see Colt, your team has a great risk of being exposed. It will be like trying to run a gauntlet, surrounded on both sides. I'll be at the NSA station in Bahrain while you are on the mission. It's right on the grounds of the U.S. Navy base there."

The General adds, "Our best analysis, Colt, is they will launch smaller craft from the yachts and just breach them up onto the beach. This would provide them with the most secrecy."

"Commander, you and your team will leave D.C. August 12th. You will meet up with Ms. Sanders and her group in Bahrain for final briefing and any changes to the plan. You will then be taken from the base by chopper to the USS Minnesota, SSN-783. It will be in the area and rigged with two SWCS vehicles, specifically for this mission. We don't want your team even THINKING about jumping, or parachuting, into the water in this crowded area of the Persian Gulf. This is why you'll use new submersibles to arrive at the surveillance area. We are in super stealth mode. Is that clear?"

"Yes sir, Admiral." Colt looks around at everyone before speaking further. "I get it. This looks like a fairly simple mission---on paper. But, the reality is, it will be a challenge to remain undetected. We will be in surveillance-mode only, and unless we are engaged, that is all we will be doing."

The General says goodbye, "We'll be in touch with you and your team at all times, Colt. Good luck."

"Thank you, sir. Thanks to everyone. Lori…Thanks for the Intel. It is spot on as usual." Colt looks directly at the Admiral, "Sir?"

"Yes, Colt?"

"May I take a few of my men from STA?"

The Admiral looks smugly at Commander Colt Jackson, "I knew you would ask this question, Commander. I have already authorized you to bring in three people from SEAL Team America, in addition to yourself. It will undoubtedly ruffle a few feathers at SEAL Team 3, but I have other things for the remaining platoons to do. They won't be sitting around, I can promise you! This will give you a total of twenty-four men for the mission.

"Thank you, sir! That will do it. A platoon from SEAL Team 3 and four men from STA, some of their skills will complement each other. I know you always have the best in mind for all of us. It will also make the partitioning of the surveillance shifts easier to manage."

Colt Jackson stands and leaves. Lori Sanders, General Shane Maxwell and Admiral Steven Whistler remain behind to go over a few other details.

CHAPTER 5

SEAGLASS & SEALSKINS

In Bahrain at the headquarters of the 5th Fleet and the U.S. Naval base, it is Saturday, August 13. Colt and SEAL Team 3 wake up after landing at 2:00AM local time, after a grueling 14-hour flight. They had gone wheels up when they departed Washington D.C. at 6:00AM on the 12th. Today, they have a 12:00 noon briefing with the NSA on the base.

Colt turns to his SEALs, "OK, men. Everybody's moving a little slowly this morning. Let's grab some chow and head to the briefing center."

Lieutenant Cody Custer, call name "Tiger Fox," has been with Colt Jackson for quite some time. He is an outstanding leader and great SEAL. He was wounded and shot twice by Priest Assassins on February 20, 2015, when the assassins were trying to murder his Commander---Colt Jackson. He is fully recovered. "Sounds good to me sir. I'm starving. Eating MRE's* during a 14-hour flight isn't exactly my idea of 1st Class!"

"The C-130 wasn't the Taj Majal, that's for sure," Colt retorts.

Chief Petty Officer Shane McCloud chips in, "Well, sir, I guess that's true, but it sure beats the rocks of Afghanistan, don't it?"

Colt Jackson and his SEALs laugh. "You've got that right, Chief! How quickly we forget..."

Chief Petty Officer Shane Duffy remembers only too well and quickly agrees, "Aye to that, sir, aye to that."

At twelve noon, the SEALs shuffle into the NSA briefing room. Attending the meeting are Lori Sanders, head of NSA PRISM Internet Operations, Michael Washington, CIA Station Chief---London and a twenty-four-man platoon from SEAL Team 3, led by Commander Colt Jackson. Due to the covert nature of this mission, there was no need to deploy any more men from the SEAL Team.

Lori Sanders begins the meeting. "Here is the map and layout of The

* Meals-Ready-To-Eat

Pearl. You can clearly see the small island where the Marsa Malaz calls home. You can also see there is only one approach from the Persian Gulf and only one approach from D'oha---the lone road leading onto the property from the Venetian District. For those of you who have not been to D'oha, or The Pearl, it is home to some of the wealthiest people on earth. They have spared no expense in replicating the best of the very best---including Venice! Over-water walkways, verandas and multi-million dollar villas surround the hotel grounds."

Michael Washington picks up from where Lori leaves off, "I'm along for the ride in this briefing because we have tailed a number of suspected Middle East terrorists into Qatar. Some of the CIA's best operatives have been exactly where you men are going in a couple of days. We know the water, the islands, the architecture and have all the blue prints for the hotel. Here they are." The Power Point presentation pops up the hotel's blueprints.

Commander Jackson says, "Great to see you, Michael and Lori. Michael, where are the likely places the meeting will take place? Do we have any information on the building---special rooms, secret tunnels, underwater passageways, etc.??"

"Great question, Commander. It turns out the Marsa Malaz was constructed on behalf of the Emir, himself. This fact has been kept secret from the general public. They have purposely limited the number of egress and ingress points to ensure security can be kept extremely tight for the protection of his eminence."

Lori continues, "Michael will provide you with the logical locations for the meeting, however, the NSA has never been able to penetrate one part of the hotel." Lori clicks for the next slide. She uses a laser light to focus attention on the Grand Ballroom.

"We have attempted to use SKYBAT to eavesdrop on the Grand Ballroom and its connected VIP suite. There is a dearth of SIGINT available from these two locations. We have attempted many times during events, when they were packed with people. We can't even pick up a cell phone signal out of these rooms."

Michael Washington concurs, "Lori has it nailed. Anything sensitive will very likely be done within the confines of this room. The rest of the hotel can be virtually bugged. It seems the Emir has placed a protective, electronic drape around the ballroom."

Lieutenant Cody Custer has a question. "Mr. Washington, or Ms. Sanders, do we have any intelligence on any of their defensive mechanisms which might protect the property? You know, intruder alert systems, patrolling guards on foot, patrolling guards on boat, or patrolling guards on jet skis. Are you aware of any drones, or infrared detection capabilities they might have at the Marsa Malaz?"

Michael Washington replies, "It is all state-of-the-art. Electronic surveillance, infrared automatically turns on at dusk and there are roaming, armed patrols are on the grounds, twenty-four hours a day."

Colt Jackson jumps in with the opportunity to enlighten his men. "Men, we have been given some new exo-skeleton, all-weather suits, officially called 'Warrior Web.' They automatically adjust temperature, of the outside coatings on the suit's neoprene rubber, to the water's temperature. But, if you're on land, the coatings also adjust to the ambient temperature of the outside air---either when you pop-up out of the water, or if you have to make a beach landing for inward Intel. What a breakthrough by DARPA. It makes us virtually invisible on the infrared spectrum."

Chief Petty Officer Shane Duffy says, "I was wondering what that coating material was for!"

Colt continues, "Sorry you weren't briefed any sooner Chief---none of us were, really. I received word from Admiral Whistler he was going to send along some special 'gifts.' He briefed me via SAT phone only this morning. The Warrior Web suit also provides significant joint and muscle support. It reapplies and multiplies the body's energy for stronger movement."

Michael Washington says, "Since your men were down on the farm and became a true partner of the intelligence agencies, the decision was made--‑above all of our pay grades, to make these suits available to you for this mission. They are perfect for your mission's use and deployment. The marine-version of these suits has been renamed and called, appropriately, "Sealskins."

A couple of the Petty Officers on the SEAL Team, shout out "HOOYAH! COOL!"

Lori Sanders continues the meeting, "As all of you have already been briefed by Commander Jackson, the Q-Warrior headgear is integrated into the Sealskins for water use. The communication capabilities are powerful. They allow personnel tracking, hostile enemy detection and include incredible night-vision capability, with HD color feeds. Information, video and audio are broadcast via secure, two-way, high-speed channel."

Colt says, "We have twenty-four men to cover the hotel---starting in only three days. This means we must deploy during the evening of the 15th." Colt looks around the room at his men and pauses as he calls out each man's name:

"Lieutenant Cody Custer, call sign 'Tiger Fox;'

Chief Ryan Duffy, call sign 'White Shark;'

Chief Spider Sinclair, call sign 'Black Widow;'

Master Chief Bull Sherman, call sign 'Sierra Hawk;'

Ensign Dizzy Lee, call sign 'Stone Fish;'

and I, call sign 'Papa Bear,' will head up six different teams. We will deploy in four-hour shifts, AROUND THE CLOCK, starting at the stroke of midnight, on the morning of the 16th.

Lori Sanders assures the men, "The NSA center here on Bahrain, will be tracking everything. The USS Minnesota will also be offshore, standing by, for support."

Colt adds some more good news for the SEALs, with respect to new toys made available to them for this mission. "We have two brand new 'SWCS' (pronounced "SWIKS")---Mark 10 models. They are being newly deployed for this mission. These Shallow Water Combat Submersibles are vehicles like you've never ridden on, before. They are extremely stealthy and will automatically camouflage themselves with the sea floor and surroundings---no matter what they are. They were developed at the direction of the U.S. Special Operations Command (USSOCOM). When you complete your four-hour surveillance shift, the next team will arrive to take your place, like a tag team. You'll use your Mark 10s and ride back underwater to the submarine and rest. One SWCS will be left near the surveillance team and the other, if it isn't ferrying SEALs back and forth, will be tethered to the sub. Each four-man team will go back out twenty hours later, each day, until the meeting is over on the 19th, or sooner---if it breaks up before then. Lieutenant Custer, brief them on our other gear requirements."

Lieutenant Cody Custer begins, "Gentlemen, the area all around Qatar is extremely SHALLOW---making it another exceptionally difficult factor working against us. Thus, we will be using our LAR V Draeger Rebreathers---the MK 25s."

The LAR V Draeger rebreather, also called the MK 25, is a closed circuit SCUBA device. It runs on 100% oxygen and a human being's expelled breath is recycled into a closed circuit where the carbon-dioxide is scrubbed out. There are no bubbles and it is perfect for Colt Jackson's surveillance team. They will be in very close proximity to the hotel and beach. The extremely shallow water in the Persian Gulf makes the rebreather a perfect fit. MK 25s, unlike SCUBA gear, are worn on the front of the SEAL and are actually smaller than a traditional SCUBA tank.

"As you know, they eliminate the carbon dioxide out of every exhaled breath. Just to give you an idea about HOW shallow it is around Qatar, even five miles out in the Persian Gulf from the hotel, the water is ONLY about thirteen feet deep and the natural channel approaching the hotel is only three! Of course, the dredged channels are 36-40 feet, and the marine entertainment areas around the hotel are a bit deeper."

Chief Spider Sinclair has a question. "Sir? Won't this make us stand out like sore thumbs in the day time? I mean our dark suits will stand out against the light of day, big time."

Colt Jackson intercedes, "If I may Lieutenant Custer?" The Lieutenant nods.

"This is the last surprise Admiral Whistler sent us. The Sealskins also have built into them, like the Mark 10s, the ability to autonomously camouflage us---in other words, it has an active camouflage system for virtually any background, no matter where you are. It is straight out of Star Wars. The cloaking technology was invented by some brilliant Canadians. The strange coating you see on the Sealskins actually bends light AROUND you---causing you to be virtually invisible in the water---or on the shore."

Lieutenant Cody Custer looks at his Commander with surprise in his eyes, "Are you kidding us, Papa Bear?"

"I kid you not, Tiger Fox. This is super-top secret. No one knows in the SEALs, other than this group and the top brass. The technology was originally developed for the Army, but has been adapted for aquatic missions."

Chief Petty Officer Shane McCloud chimes in, "I saw a video on cloaking capabilities being applied to the U.S. Army---I'm glad they were thinking of us, too!"

Colt Jackson looks out and eyeballs one of his men, "Ensign Dizzy Lee?"

"Yes, sir?"

"Stone Fish---would you like to demonstrate the Sealskin capabilities for the group?"

Ensign Dizzy Lee gets excited and springs up, "Sir, yes sir!" He puts on the Sealskin suit.

Colt Jackson orders, "Now, Stone Fish, if you would please go up to the screen where Ms. Sanders and Mr. Washington have been giving us the presentation." Stone Fish saunters up in front of the video screen and faces the platoon.

Chief Petty Officer Shane McCloud complains, "But, Commander---I can still see him as plain as day, how is this going to help us?"

Colt Jackson now orders, "Now, Stone Fish---on your right wrist there is a command wrist watch. Press the two square buttons on the opposite side of the gauge--from left to right, in sequence."

Ensign Dizzy Lee replies, "Yes, sir!" He dutifully punches the two buttons. Immediately, he disappears into the video screen with the upper half of his body, while the lower half of his body has turned into the cinder blocks of the wall behind the viewing screen.

Chief Petty Officer Shane McCloud, "Would you look at that. Shit!"

Chief Petty Officer Shane Duffy, "I'll be damned, unbelievable."

Lori Sanders says, "Merciful heavens, THAT is surreal. And it works underwater, Colt?"

Colt responds, "I've been told it has gone through exhaustive testing at DARPA and the Office of Naval Research ("ONR"). It is an amazing addition to our capabilities. Now, men, get some rest, but you need to adjust your sleep schedules in accordance with your shifts. Have a little R&R for the rest of the day, but tomorrow meet me at the SEAL briefing center at 0800."

SEAL Team 3 responds, "HOOYAH, SIR!"

The meeting breaks up and the SEALs depart to explore Bahrain. Colt Jackson stays behind to talk with Michael Washington and Lori Sanders.

Colt directs his next question to both Michael and Lori, "Do you want to get together for an early dinner in a couple of hours, Lori and Michael?"

Michael replies, "I'd love to, Colt. Unfortunately, I am heading to the South Sudan. The newest nation on earth needs our help. They captured some of the Somali Jihadist group, al-Shabab, there. They had some advanced weaponry and we want to know where it came from. Its leader, Ahmad Abdi Godane, maintains a tight grip over the group and has formalized its alliance with al-Qaeda. They have been chased out of the larger cities in Somalia, but they are expanding their influence in the region. I have to interrogate them."

"I'm sorry to hear that, Michael, but I know they have the right man for the job. Be careful…"

Michael replies, "You too, Colt!"

Lori Sanders is secretly overjoyed to be able to have a meal with Colt, alone. "What a shame, Michael. It would have been lovely to catch up on the news from London over dinner. Since moving back to the States, I miss my apartment there. There is something about SOHO in London. I love to walk through the theatre district and then just head out to dinner---with a million choices within walking distance."

"I know, Lori. London definitely grows on you. I love it there, as well. Anyway, I am being flown out as soon as we're done here. So, until next time…"

Colt says goodbye, "Thanks for everything, Michael. We appreciate the deep background and debrief you provided us, as usual!"

Michael Washington also says his goodbyes, "Colt, Lori, take care." He walks away.

Lori Sanders says mischievously, "Now, you said something about a couple of hours from now getting a bite to eat?" She smiles and as they stand outside, the sunshine seems to twinkle off her light, golden brown hair. The blond streaks come to life, as if angels with lit wings were running through her hair.

Colt Jackson is gulping, while looking intently into the bright, green eyes smiling back at him. "I sure did! I was wondering if you like Japanese food? There is an amazing restaurant here called Maki. It is right next to the Moda Mall, one of the most exclusive shopping centers in the world. It's supposed to be amazing."

"I love Japanese food, Colt. It sounds perfect! What time?"

"Can you be ready in an hour?" Colt smiles, hoping to pick her up, early.

"I'll be ready. I'll just file a few things and freshen up. See you in a little bit."

"Great, I'll pick you up at your hotel---are you at the Navy Gateway Inns and Suites? Just a wild guess, but government travel on a military base and all..." Colt smiles and winks at her.

"That's exactly where I'm at. It's not the Ritz, that's for sure, but I did get upgraded to a master suite. All the comforts of home! I'm in room 207. Just think, if you hadn't invited me to dinner, I would have just gone to the movie theater attached to the hotel!" She smiles and winks back at Colt.

"OK, then, I'll pick you up in a little while!"

"I'll see you then."

They walk out of the briefing center and into the sunshine of the Naval Support Activities Base and depart.

CHAPTER 6

THE MAKI RESTAURANT

August 13 at 3:30PM, Bahrain Time. Colt Jackson and Lori Sanders have borrowed a base vehicle and changed into civilian, conservative clothes, in line with the 5th Fleet's Commanding Officer's orders when personnel are going off the base. They are both admiring the skyline of Bahrain, as they approach King Faisal Highway. Modern buildings and skyscrapers are everywhere. Palm trees explode upward reaching out to the brightly lit, sunshine sky.

Colt and Lori enjoy the ride through Manama. They cruise up Shaikh Isa Avenue and Al Kabeer Avenue to reach the restaurant. The World Trade Center and the Maki Restaurant are located just to the right, in front of the highway.

Colt Jackson pulls up to park the car at Maki. This fabulous restaurant is famous for its Japanese fusion food. It is on the first floor, connected to the West Tower of the World Trade Center. The Moda Mall, one of the most exclusive in the world, is also directly connected to the World Trade Center.

Lori eyeballs the structure of the tall building. She then says, "You know, a girl could get lost in THIS mall! Dior, Hermes, Louis Vuitton, Fendi, Gucci, Tiffany, Rolex, Piaget, I was looking at the brochure in the room. I think I would be overwhelmed!"

Colt starts laughing. "I can understand that. I am amazed at how much money these people have. I sat in wonder once, when I watched how a Saudi family just walked into a Hermes store in London and nearly bought the place out. It was crazy! Don't ask me why I went in--- I was actually tailing someone!"

Lori laughs and looks at him with another mischievous look, "Yeah, sure you were! You were probably there with Mary Barrister! OK, just kidding. I've seen the same thing, many times in London, during the summertime, especially. RIGHT NOW in the middle of August, London is slammed by people visiting from the Middle East. I think it is a huge relief from the heat and London does have some great shops and restaurants. The Jumeirah hotels are just packed in Knightsbridge now."

Colt says, "I'm glad this is only dinner tonight! This date could get REALLY expensive!!"

Lori pounces on this and asks questioningly, "So, Colt, ARE WE ON A DATE?"

"I've been longing to go on another date with you, Lori. Between my recovery and then being whisked off with my SEALs and SEAL Team 3 from Coronado, time passed me by. Even with dinner after our meeting at the NSA in Ft. Meade, I know we haven't had much time, together. I'm sorry."

"Don't be sorry, Colt!"

"I've felt badly. You were there for me all through my recovery, even though you were on an incredibly busy schedule."

"I tried to be, as much as I could. I know I wasn't around as much as Mary Barrister..."

Lori is probing Colt for a response from him about Mary Barrister, who got very close to Colt during Operation "End the Fed." Colt saved her life, several times, and she saved his, as well. Lori is a bit jealous and apprehensive about this relationship.

Colt laughs. "You are quicker than ever, Lori. As I told you, I have feelings for both of you. It is a bit confusing. But, being able to at least have the time to spend with someone you're interested in, is a huge step forward. At least you and I can work together---and steal some moments to find out where we might go. I was wondering when you were going to bring her up. Are we OK about this?"

"No pressure, Colt. I was just wondering about her. Is she doing all right?"

"I spoke with Mary about a week ago and she seems to be doing fine. She was heading off to Switzerland to meet some friends at a Chateau. Frankly, after everything she went through in 2015, I'm glad she could get away. It's hard for anyone to get their arms around the loss of a spouse, let alone a true civilian who had to literally run for her life. She's a very strong lady. She is a good friend."

Lori Sanders cannot resist, the intelligence analyst is all over this one. "Just friends?" she says with a questioning twinkle in her eye, as her eyebrow rises.

Colt looks at Lori and smiles. "Ahh, Lori, knock it off," Colt says with a laugh. "Let's go eat some of the best sushi on the planet!"

"Sounds delicious---all right, sorry Colt. It's just my nature and my training. I have to know EVERYTHING!"

Colt laughs, "That is what I love about you, Lori. You are spectacular---in many, many ways. Someday, maybe I'll know something you don't know, and then I can SURPRISE you!"

"Colt, you ALWAYS surprise me. I am happy to just have dinner together and I'm thrilled Michael had to go off to the Sudan!"

"Poor guy. That place is a real developing nation. I hope he is OK."

Colt Jackson and Lori Sanders get out of the car and valet it. Colt is dressed in a light, tan suit, with a white shirt, and Lori is dressed beautifully in a green pastel, printed dress, with a sparkling gold necklace around her neck. Her golden, light brown and blond-streaked hair is flowing below her shoulders and she is not wearing glasses. Lori has ditched her fake horned-rimmed glasses she wears at the NSA. Her makeup is perfect and her natural smile is highlighted by tantalizing lipstick---a subtle and sexy shade of rich pink. As they get out of the car, they see the restaurant. Maki Bahrain is attached to the World Trade Center's West Tower. Both of them look up, again, and admire the Towers' wind turbines and fabulous architecture of the twin building. It is fifty stories tall. A short walk through the West Tower lobby would bring them to Fashion Court. Lori hopes to go there after dinner.

"Oh, Colt, it's simply gorgeous!" Her bright green eyes sparkle up at Colt in the late afternoon sunshine. Colt stirs at her beauty.

"It is stunning, but Lori---it's not as beautiful as you are!"

"Ohhh, thank you, Colt! Do you think we can peak inside the mall after we eat?" She squeezes his hand.

"Absolutely!"

Colt Jackson and Lori Sanders approach the restaurant entrance. The entrance has masonried beige facade with vertical, rectangular hung smoked-glass windows prominently integrated into the walls. The prominent red-circled Maki Restaurant logo and trademarked signature welcome its customers. The modern, ground floor entrance is a delight to the eyes. They enter the restaurant.

The Maki hostess greets them with a warm smile. Colt and Lori both notice she is quite beautiful. "Konnichiwa. Welcome to Maki," the lovely Japanese hostess says.

Both Colt and Lori respond in kind, "Konnichiwa."

The Maki Hostess, Akiko, is a stunning Japanese woman, with a flashing smile, perfectly brushed, long black hair, and bright, dark eyes. She has a slender build and perfect figure. She has lived in Bahrain nearly her entire life, since her parents came here to work.

Paul, the restaurant manager at the Maki Bahrain intercepts Colt and Lori. "Excuse me, Akiko. I would like to take care of Commander Jackson and his guest, personally."

"But, of course, Mr. Paul. I had no idea." Akiko smiles, again, at Commander Jackson. Colt smiles back.

Paul tells Akiko, "Commander Jackson is a Medal of Honor winner!"

Colt didn't leave his rank, or even tell reservations he was in the military and he certainly didn't tell anyone he was an MOH recipient. With a look of surprise on his face, Colt only manages to say, "I don't really like to talk about that."

"Again, pardon me, Commander. The Vice Admiral John Stiller, Commander of the 5th Fleet, asked me to take care of you this evening. It is not every day the Maki receives a Medal of Honor recipient."

Colt Jackson is stunned his identity is already known, locally.

"W-h-y, thank you. What is your name?"

"I am Paul, the discreet Restaurant Manager of the Maki Bahrain."

Lori Sanders comes to the aid of Colt, who she knows like a book. She can see he is shaken up by the knowledge the manager possessed. "Why thank you for your kind hospitality, Paul."

Snapping out of his shock and taking Lori's lead in being gracious, "Yes, thank you very much Paul. We have heard a great deal about this wonderful restaurant of yours!"

Akiko observes the handsome, rugged Colt Jackson with even greater interest from this point on. She smiles warmly at Colt and wishes him a good evening, bowing her head slightly to him.

Paul shows them the way to their specially reserved table, "Please, this way to our VIP table, upstairs. Only the best of the best for you and your date, on this fine day Commander. The Admiral told me to tell you it's on him and he's sorry he can't meet up with you this trip."

"I'll make sure I thank the Admiral. He goes back a long way with a dear friend of mine and I've known him since my Academy days. He's a very good man."

"In any event, we will now leave you in peace. Have a wonderful dining experience. I suggest the specialty items and signature dishes on the menu. They are truly to die for."

Lori answers, "Well, let's hope not..." She laughs and winks at Colt.

Paul departs. Colt and Lori order from the menu.

"What looks good, Lori?"

"EVERYTHING!! I mean look at the list. Where do you even begin! Gindara Shitake Black Cod, Creamy Lobster Tail, Kung Pao Smoked Salmon...YUM!"

Colt is looking at the hors d'oeuvres. "Their starters are also looking interesting. I love tempura prawns, they have a special called Mowaffak Maki. The prawns have avocado, mango, plum wrap and I heard whispers their sashimi is unreal. But, the Mongolian Beef might also be terrific."

Lori is getting excited by the menu. "Let's order a bunch of everything! Can we have some Tom Yum shots? They have lemongrass, kaffir lime leaves, galangal, mushrooms, pounded Thai chili peppers, fish sauce, hot and sour and herbs galore, it sounds amazing. I just realized---I'm starving! Do you think this feast will be OK with the Admiral?"

"I think the Admiral has a pretty good entertainment budget. He can handle it. In fact, the last time he was in Washington D.C., he got together with me and Admiral Whistler. I got them both to meet me at the Peking Gourmet in Fairfax. I got there first---bribing the owner to not accept their money and to make sure I got the bill!"

"You're incorrigible, Colt!" They both laugh, together.

"They are my senior officers, but we go back a long way and have always hit it off. We all were on leave, so we had a pretty good night of it. I think between us we polished off two of the ducks, a couple of bottles of the finest red wine and some amazing starters. I can still taste the delicious green tea ice cream in my mouth."

"The Peking Gourmet...isn't that where all the Presidents have gone for their duck?"

"That's right, Lori. When you walk in, you see pictures of pretty much anyone famous who has entered its doors---and the pictures include just about everyone---including the Presidents."

"You know, I've never been there. Maybe I could take YOU there sometime. I love Peking Duck."

"That would be another great date!" Colt and Lori look at each other for a moment and put their order in.

They chat over dinner, finally finding time to relax, together. Colt is taken with Lori and Lori is all eyes on Colt. She feels tingly inside and couldn't be happier.

Colt exhales, "I'm so stuffed, it would choke a horse! The Dragon Shrimp and Smoked Braised Ribs were out of this world. What was your favorite?"

"The Smoked Kung Pao Salmon. I love how their presentation works. The clear, glass bowl over the wonderful smokiness---do you think it was dry ice?"

"I don't know, but it sure was delicious. How about that Chocolate bomb we had for dessert?"

"Colt, you KNOW chocolate is my weakness. It was scrumptious!" Lori looks around the restaurant and spots the ladies room. "Would you excuse me? I have to use the ladies room."

"Of course! In fact, I might run to the men's room, as well." Colt Jackson and Lori Sanders leave their table. Once they get up, servers immediately tidy everything up and make the table perfect for their return.

Colt looks at his watch. It is now 5:00PM. He walks into the bathroom to relieve himself and is standing at a urinal. The fixtures inside the bathroom remind him of something from an Asian Emperor's palace. It is beautiful, even for a bathroom.

Colt is peeing, with his back to the door and the wash basins. He hears a door open and continues to go about his business at the urinal. All of a sudden, Colt's spider senses begin to tingle, as he hears a knife get pulled out of a sheath! Colt doesn't have time to even zip up his trousers. He whirls and turns---catching the would-be assassin off guard at the sight of Colt's penis protruding from his pants!

The obviously Middle Eastern assassin is holding a military, black, long-edged knife. He menacingly is coming at Colt.

"You American Pig...Medal of Honor my ASS! You die tonight!" He swings the knife down on Colt and Colt blocks it with a sharp, left-handed, forearm thrust to the assassin's right arm, catching him perfectly at the wrist. The knife clanks to the floor, but has nicked Colt across the top of his hand. He is bleeding.

Colt immediately is in a bear hug with the assassin. The assassin seems to be well trained, as he upward thrusts his knee into Colt's protruding penis. This clobbers Colt, knocking him nearly senseless, but simultaneously Colt swings his right fist directly into the Adam's apple of the assassin. The assassin buckles over and can't breathe. Colt ignores the agony of the shot to the balls and uppercuts the would-be assassin to the jaw. The assassin is knocked out cold. Colt takes deep breaths and regains his composure and then assesses the damage he has taken. He has blood running down his hand, but he doesn't see any that got on his clothes. "*So far, so good,*" he thinks.

He grabs the assassin and quickly checks his ID. "Muhammad Al-Karim" is written on his Saudi Arabian driver's license. He checks his pulse. The assassin is alive.

Colt looks down at the unconscious attempted murderer and says, "Tough night for you, huh, Muhammad?"

Colt grabs the brown, light-skinned Muslim and places him on the toilet. He takes large, linen napkins from the sink and stuffs them into Muhammad's mouth. He uses the napkins to also tie his hands behind his back, and his feet together. He then ties linen napkins around his neck, binding the assassin to the stainless steel pipes, attached to the back of the toilet.

"That won't hold you for long, Muhammad, but hopefully you won't want to get caught in here with your knife," Colt says to the silent figure of the would-be assassin.

Colt does not touch Muhammad's knife, but rinses it off in the sink and then dries it off with a linen napkin. He then places the killer's knife near the assassin inside the toilet.

Colt thinks, "*The last thing I need right now is an international event in one of the most famous restaurants in Bahrain!*" Then he says to the assassin, "See you later, chump."

Colt closes the toilet door and leaves the men's room after washing the blood off his hands. Fortunately, no one needs to use the bathroom in the late afternoon at Maki. This was their slow time, thank God. Colt takes a detour to the Maki bar.

It is now 5:15PM. The very hip and cool Maki bar uses the same coordinated-design scheme as the outside of the restaurant. Its light-beige color and design at its base, reaches up to meet a rich, dark-serving top which comfortably can serve a small crowd. The connected lounge, with its ever-playing high-definition TVs, easily handles overflow.

It's not crowded and Colt's approach at the bar has drawn the Maki bartender's attention. "Can I help you sir?"

"May I please have a paper napkin?"

"Of course, here you go."

Colt rips the paper napkin and uses it to create a makeshift bandage for his wound.

Colt leaves the bar and heads back to Lori. She will be wondering why he was delayed. He decides to tell her something didn't agree with him---not quite a lie, but it was the best he could do. He would keep his left hand out of her site.

Colt turns to the bartender and says, "Thank you very much!"

Colt Jackson returns to the table and takes his seat. Lori Sanders is waiting patiently for him.

"Is everything OK, Colt? I was getting worried."

"I hate to say it, but something didn't agree with me. I think I must have eaten too much!"

"You poor thing! Are you going to be all right?"

"Yes, no worries, Lori. It was just a passing moment, I feel better already. I am looking forward to the rest of our evening. Shall we get the check?"

"Sounds good, Colt. Thank you so very much, I'm sorry you didn't feel well, but it was delicious!"

"Don't thank me, Lori. We'll have to thank the Admiral. AND it WAS delicious!"

Colt forces a smile and realizes the blood is soaking through the napkin. Colt spots the waiter and asks for the bill. "May we have the bill please?" The waiter says to Colt, "There is no bill. Mr. Paul said this was already taken care of."

"Well, thank you very much!" Colt looks at Lori and directs his conversation back to her. "Shall we take off?"

"By all means! Maybe we can check out the Moda Mall. I'm dying to see it."

Colt laughs. "Let's do it!" Colt will not let the bathroom incident with Muhammad Al-Karim, get in the way of this evening.

They head out the exit of the restaurant. Akiko says goodbye, somewhat surprised to see them leave. She smiles and warmly takes Colt's hands, as he passes by, "Ohhh, Commander! Thank you very much! Hopefully, we will see you again, sometime. Goodbye."

"Thank you very much, Akiko. Everything was outstanding. Hopefully, we'll be back soon." Colt smiles warmly at her.

Lori gives the smiling, vivacious Asian a wicked stare. Lori hates it when women flirt with Colt.

After they leave the restaurant, Akiko picks up her mobile phone. She speaks into it, "Come on, answer Muhammad! Pick up!!" There is no answer. Akiko places another phone call.

"Hello, Hassan? Muhammad is not answering his phone and Commander Jackson and his date just left the restaurant. Can you get someone into the men's room and find out what is going on with Muhammad?"

"I'll be there in a minute. I'm just around the corner."

Colt Jackson and Lori Sanders are walking through the lobby of the West Tower of the World Trade Center. They enter the Moda Mall, which is attached to it and find themselves in the middle of the Fashion Court. Lori is in seventh heaven.

Moda Mall is located on the ground floor of the Bahrain World Trade Center. This center is an iconic location, and is the Kingdom of Bahrain's first ultra, high-end shopping complex. It is host to a vast array of leading luxury retailers in Bahrain.

Lori looks around, "It's BEAUTIFUL, COLT!! Have you ever seen such luxury in your entire life! The light-blue lighting, the stunning architecture, THE SHOPS!! Look at them." Lori is completely mesmerized by the wealth and sheer number of the world's most prestigious shops in this mall.

Colt looks around, too. "It is really something! Do you think I can find a pair of Levi's here?"

Colt laughs and Lori joins him. She has moved to his left side, as she immediately spots the Louis Vuitton shop.

"Louis Vuitton! Would you mind if I stopped in there to check out their purses! I just LOVE THEM. They ARE expensive, but sometimes a girl's got to do, what a girl's got to do!"

"I can window shopping with you, all day long, Lori. Let's go!" Colt forces a smile, his cut is hurting a bit.

Lori Sanders glances up at Colt's face and then she looks at him carefully. Finally, Lori notices something is not quite right with Colt. She sees something in his smile. It isn't as relaxed as it was during dinner and it looks like he twitched. She stops in the middle of the Fashion Court and looks him right in the eyes.

"Don't lie to me, Colt. ARE YOU ALL RIGHT?"

Colt struggles to come up with words, "Uhhh…"

Lori looks at him, up and down. She suddenly grabs his left hand. She feels the napkin and something wet on his hand.

"Oh my goodness, Colt! YOU'RE BLEEDING!!"

"It's only a scratch, Lori. I nicked it in the men's room."

"Let me see it!" She lifts up his hand and pulls up the napkin. It reveals a clean cut---it wasn't a nick.

"COLT! WHAT IS GOING ON!! WHAT HAPPENED TO YOUR HAND!" She inspects him more carefully, examining his entire body and his light, beige suit. There are barely noticeable blood spots on his right arm and his knuckles appear to be slightly swollen.

She nearly screams at him, "COLT JACKSON!! You tell me right now what happened, or I swear I will get a taxi and go home. I can handle the truth, but I can't handle being in the dark! Please TELL me what is going on!"

"I'm sorry, Lori. I was just trying to give us a good evening and not ruin it by a bit of unpleasantness. I was attacked in the men's room."

"What?? What happened?"

Colt explains how an assassin snuck up behind him and how he disabled him, without killing "Muhammad".

"He tried to get the drop on me---while I was peeing---can you believe that? Anyway, something didn't seem right when the door opened and no one walked up to a stall, or the urinal. I guess my spider sense went off. I turned just in time to block the kill-thrust, but got nicked in the process. I really just need a good bandage."

"Where is Muhammad?"

Colt starts laughing. "I tied him up with linens, both his hands and feet. Then I tied his neck around the toilet's fixtures. He'll get loose after he wakes up. But, as a Muslim, he'll feel disgusting. Hah! Serves him right."

"You didn't kill him?"

"What and ruin a perfectly good evening?" He laughs, again. "Seriously, Lori, this is probably our one chance, for a while, to spend a nice evening together. If it is any consolation, I tried my best to not get dirty..."

"You are truly something, Colt. Was this random? It seems unusual you would be targeted. There are many Americans who come in here and there hasn't been an incident."

"I'm not sure, but when we get back, I'll report in to Admiral Stiller and find out what he knows about 'Paul.'"

"That sounds like a good idea. What about Akiko?"

"I don't know. She seemed nice enough." Colt raises his left eyebrow up and down, teasing Lori, because Akiko was an Asian beauty.
"Colt! Enough."

"Come on, Lori, I only had eyes for you! You look simply gorgeous tonight. Besides, there aren't too many Japanese tied up with the Muslims. Japan banned Muslims from being in their country. Muhammad did call me an American pig and he knew about my MOH, though. Paul may be a likely suspect. In any event, it's over now. They won't try and touch me in public. Let's enjoy our evening."

"OK, but, let's get to security and find you a bandage and some ointment." They head off to security.

CHAPTER 7

GHOSTSHIP, SEAHORSE & TIGERS

On August 15 at 9PM in Bahrain, the twenty-four men on Colt Jackson's SEAL Team take off in the MH60S Seahawk. It is a short ride to the USS Minnesota, call sign "Ghostship." It is one of the newest Virginia-class nuclear attack submarines in the U.S. Navy.

It was placed on active service and launched on September 7, 2013 at Newport News, Virginia. The submarine was built to keep both GD Electric Boat and Newport News Shipbuilding and Drydock Company, the only two U.S. shipyards capable of building nuclear-powered vessels, in the submarine-building business. Newport News builds the stern, habitability, torpedo room, sail and bow, while Electric Boat builds the engine and control rooms. The two companies alternate work on the nuclear reactor, final assembly and testing of the submarine.

Night & Day, Navy SEALs Fast Rope Down To Submarines

Public domain images. Work products of U.S. Government.

When it emerges from the Persian Gulf, a nuclear attack submarine is a wonder to behold from the door of a hovering Seahawk helicopter. The SEALs admire the sub's dark, sleek body, but they waste no time. They fast rope onto the hull of the Minnesota and immediately head below deck. The whole operation takes only minutes, despite the fact it is nighttime.

Colt looks at his watch, set to military time. Sunset has already taken place at 18:08, nearly three hours earlier. He is the last SEAL to go down the hatch of the sub. No sooner is the hatch closed, the USS Minnesota, call sign "Ghostship," submerges and disappears into the dark recesses of the Persian Gulf.

The US Navy will neither confirm, nor deny, any US submarine's top speed, however, the USS Minnesota is rumored to able to travel at speeds in excess of 50 knots. The wall-thickness and diameter of Minnesota's cold-rolled, steel and titanium inner pressure hull, allows the submarine's design to include a safe-diving depth of 1600 feet. Rumor also has it she has gone much deeper, nearly double this amount, before reaching her collapse depth---the point where the hull would be crushed under the pressure of the ocean.

The Minnesota is designed with SEAWOLF-level acoustic quietness for stealth, as well as acoustic tile cladding for active acoustic signal absorption. The sub also has a 9-man lock-out chamber, which can be used with Advanced SEAL Delivery Systems ("ASDS"). These are mini-submarines capable of "dry" delivery by a SEAL team. The internal torpedo magazine space can also be modified for SEALs, especially. It can accommodate up to 40 SEAL team-members and their equipment.

Submarine Side Door Lock Out Chamber

Public domain image. Work product of U.S. Government.

SEAL Team 3 only has 24 men on board. The Minnesota handles them, like a piece of cake.

The USS Minnesota is led by Commander Brian Tenclaw, a Minnesota native. The most advanced submarine in the Virginia class was christened by a contingent of Minnesota's leading citizens, who were at its commissioning.

Commander Brian Tenclaw greets Commander Colt Jackson, as he comes down the ladder on the inside of his "boat." "Welcome aboard, Commander!"

"Thank you, Commander. It's good to be here. Thanks for hosting us the next few days. Your boat is beautiful."

"No problem and thanks, we love her. I was briefed on what's going on. Let me show you to your quarters and the lock-out chamber. It will hold up to nine men at a time."

"We're going to be running on four-man shifts, so that works great."

"Excellent. Consider this your home, Commander Jackson. Let me know if there is anything you need. The Executive Officer, Robert Webber, will see to your needs before you deploy on the "Seahorse"---by the way, that's what we nicknamed those new SWIK's. They look pretty cool!"

"I'm sure they are nothing compared to what you have onboard this baby, Commander!"

"I'll give you a personal tour after you are settled in, or whenever you have your rest period. I know you have to launch immediately."

"I'll take you up on that, Commander. I actually walked on the hull of this ship when it was being constructed at Newport News. What a boat!"

"Wait until you see what's under the hull! It is amazing."

"Thanks, again, Commander."

"Don't think about it. No problem."

At 9:55PM the evening of the 15th of August, Colt Jackson leads the first group out. Upon arrival aboard the sub, they stow their gear and put on their Sealskins. They also strap on their MK 25 LAR V Draeger rebreathers. The Q-Warrior, robo-vision headgear from the NSA is also put in place. Cleverly, the SEALs nicknamed the Q-Warrior head gear, "Seaglass."

They will use the Seaglass integrated night vision, and high-definition video and audio feeds, to record the comings and goings of the Islamic

State meeting. They are also armed to the teeth. The team waterproofed their HK-416s with condoms stretched over their 5.56 mm muzzles. They are ready to fire, if needed. If the SEALs need to pop up and shoot, they will be ready. They also have grenades and HK 45C handguns, with hollow-point rounds. The .45 caliber handguns will stop just about anything in its path, in a close fight. The SEAL team has fire suppressors for every weapon, except one.

It is impossible to put a suppressor on a bang stick. Bang sticks are made to keep ocean predators at bay. Silence and stealth are critical to the team's survival and success, even under the depths of the ocean. But there is no masking the sound of an explosive bang stick going off into the head of a potentially deadly predator.

At 10:00PM, Colt and his three SEALs, Master Chief Jack Flintlock, Petty Officer Duke Moran and Petty Officer Paul Westwood swim out of the chamber and approach one of the Seahorses, attached to the Minnesota, on the top of her hull. It is in its shelter.

The Seahorse's top speed under water, is 15 knots, a breakthrough for this type of delivery vehicle. It was made possible with new, advanced composite materials which make the hull extremely slippery. When you rub it, it feels surreally smooth. The only way to describe how slippery it is would be to imagine trying to walk over a clear ice sheet---with a pair of fancy leather shoes. This new material, combined with more powerful engines and revolutionary batteries, allows a Seahorse to be on mission for longer than twelve hours. It only requires a single driver and provides the driver with revolutionary hydrodynamic optimization and dynamic propeller capabilities.

The auxiliary open-circuit (e.g. SCUBA) breathing system supports up to six divers for 1.5 hours at 25 FSW (Feet of Salt Water) and 10.5 hours at 15 FSW before recharging is required.

The trip to the Marsa Malaz Hotel from the Ghostship Minnesota is twenty miles away. With the Seahorses' top speed of 17.2 miles per hour, or 15 knots, the round trip will take slightly more than two hours---70 minutes each way. In total, each SEAL surveillance unit will be on their rotation for slightly longer than six hours, albeit with two of them, joy-riding aboard the Seahorse.

The Minnesota couldn't get any closer than twenty miles to D'oha and

The Pearl because it has a draft of nearly 34 feet. The depth of the Persian Gulf---even twenty miles offshore, directly east of The Pearl and D'oha, is only 55-60 feet. With a $2 Billion submarine at stake, this is as close as Commander Tenclaw was willing to get.

The Seahorse's secure, tethered underwater communications systems link Diver-to-Diver, Diver-to-Seahorse, and Seahorse-to-Sub with constant communication capability. The Seaglass headgear being worn by the SEALs, is also integrated into the Seahorse communication system. Everything the SEALs record with their Seaglass will be relayed and watched in real-time by the Minnesota, the NSA and the command team.

Each Seahorse is capable of engaging in a 12-hour mission, provided the SEALs have extra oxygen, or rebreathers, on board. The one-hour hop to The Pearl makes the business of transporting SEAL surveillance units to the hotel an easy job for this revolutionary vehicle.

After emerging from the submarine's lock-out chamber, deep beneath the Persian Gulf, Colt Jackson and his team are amazed with the strange, luminescent coating on the Seahorse. It changes colors and patterns depending on the person's viewing angle. It also makes the Seahorse virtually invisible in the water. The craft blends in perfectly with the background, in all directions, regardless of a person's visual perspective around the craft.

Colt can't control himself, "WOW! Would you look at this beautiful machine? It reminds me of a chameleon, or an Octopus which changes color, depending on the rock it's on!"

Master chief Jack Flintlock, Call sign Killer Whale, echoes his Commander's sentiment, "That's really weird! The way it blends in with the background reminds me of how flounders hide themselves against the bottom of the ocean."

Petty Officer Duke Moran, call sign "Stingray Charlie," also joins the chorus, "How cool is that!" He looks over at Petty Officer Paul Westwood, who is call sign "Ocean Dragon". They are both wizards with electronics. "Ocean Dragon, take a look at the electronics on board!"

Ocean Dragon speaks into his integrated headgear microphone, "The COMM system is amazing. We can talk SEAL-to-SEAL, SEAL-to-Seahorse, or SEAL-to-Sub in real time, Stingray Charlie."

Stingray Charlie replies, "Yeah, and this baby will do 15 knots---over 17 miles per hour underwater. That's over a 50% speedup from the old SDVs (i.e. SEAL Deliver Vehicle). Supercool."

Colt Jackson, call sign "Papa Bear" says, "All right guys, let's release her and get moving. We should be there a little after 11PM. It will give us time to reconnoiter and get in place."

Papa Bear, Killer Whale, Ocean Dragon and Stingray Charlie release the Seahorse. They take off and cruise full speed at 15 knots directly west, with their GPS coordinates set at 25 degrees, 22 minutes and 37.63 seconds north, by 51 degrees, 36 minutes and 51.18 seconds east (25° 22' 37.63" N, 51° 36' 51.18" E). This will put them just outside the approach to The Pearl, on the western edge of the Gulf.

They know they are on an extremely sensitive mission. Not only are they trying to surveil the world's leading group of terrorists, but they must travel very close to Iranian waters. The international boundary separating Iran and Qatar is set along a geodesic line set at the extreme western part of the most western portion of the boundary north of the continental shelf belonging to Qatar. It is linked to point (2) below, at an angle of 278° 14' 27" by a geodesic demarcation line.

	Latitude	Longitude
Point (2)	27° 0' 35"	51° 23' 00"
Point (3)	26° 56' 20"	51° 44' 50"
Point (4)	26° 33' 25"	52° 12' 10"
Point (5)	26° 06' 20"	52° 42' 30"
Point (6)	25° 31' 50"	53° 02' 05"

The boundary is approximately 130 nautical miles in length and runs in a northwesterly-southeasterly direction in the central portion of the Persian Gulf.

The team doesn't want to incite an international incident with Iran during this mission. As a result, their planning avoided this demarcation line. It was just another thing to plan for before the mission, when they plotted their approach path to Qatar.

The Persian Gulf is one of the most watched over places on planet earth. They knew anything the team could do to avoid trouble would be critical to achieving a successful mission.

As Colt Jackson and his SEALs cruise through the Gulf, they rely solely on the Seahorse sonar, electronic navigation and autopilot to make a dash to their GPS drop point near The Pearl. In the dark, it is difficult to see much of anything, but it is an eerie site for all of them to watch the mammals pass by,. along the way. With their night vision in their Seaglasses turned "on," the mammals' warm bodies show up, with incredible clarity. Dolphins abound, living happily among all the drilling rigs in the area. There is even enough light from the moon to make out large, Hawksbill Sea Turtles. The Hawksbill is the only species in the genus Eretmochelys.

Hawksbill Sea Turtle
Original upload by
Aquaimages at en.wikipedia
http://en.wikipedia.org/wiki/Hawksbill_sea_turtle#mediaviewer/File:3959_aquaimages.jpg

The SEALs get quite a scare, however, when a very large Dugong Dugon---a sea cow, nearly hits their vehicle. Colt quickly grabs the steering wheel of the Seahorse, which automatically disengages the auto pilot. He turns the wheel harshly to avoid a collision with the large animal.

Colt Jackson speaks over the secure COMM system, "Man, THAT WAS CLOSE! Those Dugongs are gentle, but if we had collided with them with the composite materials of the Seahorse, we would be demolished!"

Petty Officer Paul Westwood, call sign, "Ocean Dragon," replies "Holy smokes! Great move, sir!"

Dugong Dugon
Public domain image.
http://en.wikipedia.org/wiki/Dugong#mediaviewer/File:Dugong.jpg

Master Chief Jack Flintlock, call sign, "Killer Whale," says, I can see your reaction time hasn't slowed one bit, Papa Bear!"
Petty Officer Duke Moran, call sign, "Stingray Charlie," joins in, "Papa Bear, do you think our handlers will let me grab a snapshot of that close encounter for my Facebook page? We missed that sea cow by inches! The video would be wild to watch---especially lit up with the night vision!"

"Don't even ask, Stingray Charlie, don't even ask. You are being funny, right?"

"Of course, sir. Give me some credit. Posting a pic of a Dugong from the Seahorse would blow the secrecy, and location of our mission, in a nano-second!"

Ocean Dragon supports his buddy, Stingray Charlie, "But sir, you have to admit, it would be pretty cool, though."

"All right, I admit it. We're almost there. Buckle up. The approach should be OK for us---and deep enough. Before they excavated the channel and approach-path to the sequestered island containing the Marsa Malaz, at low tide, the water was only about a foot deep. Because of the planning for the big yachts and playground of the rich by the Emir, the channel and approach to the private beach and cove of the hotel should be perfect for us."

Killer Whale acknowledges, "Roger that, sir. I understand the water is extremely blue and clear in the areas where they have excavated. I hope our new suits and the active camouflage systems work!"

Stingray Charlie affirms immediately, "They sure look like it to me. Along the way, I was watching both the Seahorse and all of you guys blend into the scenery. Even with the darkness of night, there is a full moon three days from now, and the moon's bright light lit up the sea floor. The Seahorse blended right in---and when I stood up and changed the angle of my eyes watching you guys, you also blended in with the changing sea floor as it whizzed by!"

Petty Officer Paul Westwood joins in, "Yeah, me too. I couldn't resist, along the way, checking out the active camouflage. This is stuff out of the cloaking scene in Star Wars."

Killer Whale asks, "Where do you think DARPA got the idea from, anyway??"

At 11:00PM on the 15th of August, The SEALs arrive at 25° 22' 37.63" N, 51° 36' 51.18" E, where the Persian Gulf meets the entrance into D'oha and The Pearl. They are four miles directly east of the hotel and their surveillance objective.

Colt Jackson reviews their plan, "Our target is 25° 22' 37.07" N, 51° 32' 59.65" E---the beach at the Marsa Malaz for one set of eyeballs. We also need to set up surveillance at 25° 22' 37.50" N, 51° 33' 01.69" E, along the jetty, which should be good for cover, as well as at the boat dock at 25° 22' 51.60" N, 51° 32' 59.02" E. This will leave one of us free to patrol and scout. The Seahorse is fast, fully up to spec on her 15 knots/hour. We should be at our anchor zone in about 10 minutes then we'll set the Seahorse down several hundred yards away from the entrance to the lagoon of the hotel. From there, we'll snap on the Pegasus Thrusters, and we'll be there in a flash."

Petty Officer Duke Moran confirms this information with his Commander, "We sure will, sir. For short distance hops, these military grade thrusters are the bomb. The Pegasus flies at 170 feet per minute---and our hands are free."

The Pegasus Thruster is an underwater Diver Propulsion Vehicle (i.e. a "DPV"). The main difference between the Pegasus and other DPVs is the Pegasus propulsion system snaps onto the back of the diver. This leaves a Navy SEAL "hands free" underwater and propels him at high speed, just like a dolphin.

Petty Officer Paul Westwood speaks to his friend, "Yeah Stingray, the old DPVs (Diver Propulsion Vehicles) always had our hands tied up. Now, they're free. I calculate it will only take us five-and-a-half minutes to be at our surveillance points with these babies strapped to our back. They turn us into miniature submarines, with a propeller pushing us forward."

Commander Jackson reminds them of their rules of engagement, "Remember our rules of engagement, gentlemen. DO NOT ENGAGE UNLESS WE ARE ENGAGED UPON. GOT IT?"

All the SEALs reply, "HOOYAH, SIR!"

Master Chief Jack Flintlock, Killer Whale, directs his question to Commander Jackson, "Papa Bear, do we still split up as planned when we reach the lagoon, sir?"
"Aye to that, Killer Whale. Let's hit it. No time to waste. I want to be there well before midnight. Gentlemen, be very careful. We know the sophistication of ISIL and the Emir are advancing rapidly. I don't need to remind you what they would do to one of us, if we were captured."

Petty Officer Duke Moran replies, "We've seen enough heads roll, sir. We know."

Petty Officer Paul Westwood tries to put his Commander at ease, "Don't worry sir, we got this one."

The SEALs open the storage bin of the Seahorse and extract the equipment and weapons they need for the mission. Each man has his pack ready. They first put on their MK 25 LAR V Draeger rebreathers. This will prevent any bubbles going to the surface and they are much smaller than traditional SCUBA gear.

They grab their HK-416s rifles, which still have the condoms stretched over their 5.56 mm muzzles, ready to fire, if needed. They also have grenades and HK 45C handguns, with hollow point rounds, waterproofed like the 416s. They each carry a single bang stick, an unnecessary precaution in these waters, as sharks rarely attack anyone around here and would not very likely attack a group of suited-out warriors, like these SEALs.

The last thing they load up on, are hydration packs and fresh water. This team is operating at nighttime, but in the daytime when the temperature hits 110 degrees Fahrenheit, dehydration is a real issue. In an emergency, the Seahorse desalinization pump can be used if a team runs dry. A capped straw fits into a water tight seal on their water bladders, called "Camelbacks" and can fill them up in only a couple of minutes.

At 11:15:30PM, the SEALS turn off their Pegasus Thrusters and arrive, exactly five-and-a-half minutes later, at the lagoon entrance to the Marsa Malaz---just as Petty Officer Paul Westwood had calculated.

The SEAL divers are invisible for all intents and purposes. Specialized sensors allow them to track each other via their Seaglass, which outlines their teammate's location and shape, perfectly. Colt Jackson gives the order to activate the COMM links for each SEAL to enable the live camera video and audio feeds. This will relay all audio and video transmissions through the Seahorse and then on to the NSA, the USS Minnesota and the Pentagon in Washington D.C.

"OK gentlemen. Let's light up the COMM!" Papa Bear's voice is crystal clear over the secure audio COMM channel. The other three SEALs all give him a "thumbs up" sign. Immediately, the SEALs hear the slight and very faint sound of a battery coming to life, inside their Seaglass. It reminds Colt of a whale call, "beeeEEEEUH!"

Colt then tests the system out. "Papa Bear to Ghostship, do you read, over?"

Inside the USS Minnesota nuclear attack submarine, the Common Submarine Radio Room, "CSRR" comes to life at 11:16PM with Colt Jackson's voice and an image of the underwater scene before Papa Bear.

The advanced technology available in one of America's newest nuclear attack submarines is truly remarkable. It is not out of just Star Wars, but from an even different, more advanced galaxy, far, far away. The 3D-color video virtual holographic display allows Commander Brian Tenclaw and Executive Officer Robert Webber to watch, in detail, the scene before the SEALs. They can see the outline of the team through the infrared cameras in the Seaglass, as well as everything around them.

All the software programs for the Command-Control System Module, or "CCSM," on the submarine are compatible with the Joint Military Command Information System, JMCIS, which is being used at the Pentagon. An adaptation of it is also being used by the NSA. The EHF transceiver on the USS Minnesota has LINK-22, LINK-16, LINK-11 and NATO LINK-11 compatibly programmed for the Joint Tactical Information Distribution system, "JTIDS."

Adding to the submarine's formidable communications capabilities is the Enhanced Position Location Reporting System ("EPLRS"). It is a high capacity data link which provides simultaneous voice, data and situation awareness for the command. In the event of a real war, it is a self-forming, self-healing mesh network, which can automatically relay from one radio to another. The Command Data Link is used for receipt of signal-and-imagery intelligence data from remote sensors, and for the transmission that links sensor-control data to and from airborne and submarine platforms. ViaSat, BAE systems, Siemens, Northrup Grumman and several other contractors had a hand in building this other-world communication capability on the Ghostship, USS Minnesota.

While the SEALs surveil their target, the USS Minnesota surveils the Gulf, ensuring no enemy sneaks up behind them. The Ocean Surveillance Information System, OSIS, receives, processes, displays, and disseminates joint-service information regarding fixed and mobile targets on land and at sea.

Commander Tenclaw, gets on the comm, "Papa Bear, this is Ghostship, we read you and see you, loud and clear, over."

Commander Colt Jackson replies, "Ghostship, operation "Snake Eye" is underway. Over"

Meanwhile at the NSA Bahrain communication center, Lori Sanders is watching the virtual, holographic monitors record the mission.

Lori Sanders is call sign "Dolphin." "Papa Bear, this is Dolphin. How is the Seaglass?"

Colt Jackson replies, "Simply amazing. Please send my thanks to the boss."

"Will do. You guys never cease to amaze me with your selection of colorful call signs and mission names. They are brilliant and most of the time, funny. Anyway…'Snake Eyes…' interesting you selected the animal with the best thermal vision on the planet."

"We had a revelation on the way over. Hopefully, Polar Bear (the call sign for Admiral Steven Whistler) will be all right with it. We didn't speak about that detail."

"Papa Bear, you're speaking with me right now. Sounds like a perfect name. Good luck and good hunting. From here on out, all inbound COMM to the SEAL team is limited to the chain-of-command, as usual. Either myself, or Ghostship has the ball on this. Any Intel, or feedback, will go through the other encrypted COMM channel, unless otherwise ordered."

"Thanks, Polar Bear. Copy that and understood. Over and out."

"Copy that and understood. Good luck, gentlemen."

General Shane Maxwell, call sign, "Four Eyes," jumps on the SEALS live Seaglass COMM before it is shut down during the mission, "This is Four Eyes, good luck with Snake Eyes. God Speed."

Colt Jackson, Master Chief Jack Flintlock, Petty Office Duke Moran and Petty Officer Paul Westwood approach the entrance to the lagoon of the Marsa Malaz Hotel. The SEALS had read up on the Michelin three-star resort and it is supposed to be "off the charts" in the luxury department, according to people "in the know."

Petty Officer Paul Westwood, Ocean Dragon, speaks over the SEAL COMM, "This water is beautiful. During the daytime, it must be really something."

The Master Chief swims ahead of the group a bit to get a better look at the entrance to the lagoon. Killer Whale calls back to the men, "What's this? A net?"

Papa Bear comes back, "Come again, Killer Whale?"

Master Chief Jack Flintlock tells his Commander, "I've run into a net. It appears to be made of filament. Do you think it is motion sensored?"

Commander Jackson replies, "Our suits and Ghostship scan remotely for electronics in our path. I have no readings. Do any of you?"

The SEALs reply, "No, Papa Bear. Nothing."

At 11:22PM, the SEALs inspect the net and determine it must have been set up to prevent unwelcomed swimmers from entering the lagoon during the Islamic State's summit meeting. The SEALs cut a hole in the very bottom of the net to get through. There is still nothing showing up on infrared and they have only seen biologicals along the way. They swim through the hole in the netting. True to form, the Master Chief takes the lead, with both Petty Officers right behind him, swimming in formation. Colt is at the back of the pack. By 11:30, they are in the middle of the lagoon, about 500 feet from the filament net they swam through.

Suddenly, Master Chief Jack Flintlock, call sign Killer Whale, cries out.

"UGHHHHH!"

Killer Whale lets out an uncontrollable, guttural moan. A very large Hammerhead shark has collided with him. The camouflage is working, but even the sharks can't see the SEALs! The shark is as startled as the Master Chief. It swims away.

"A hammerhead hit me!"

Papa Bear, concerned, says, "A HAMMERHEAD? Are you injured? Killer Whale, ARE YOU INJURED?"

The tense voice of the CSRR control room operator from Ghostship urgently calls out, "Are you OK, Killer Whale?"

"I'm OK. I don't think it even saw me. It was as startled as I was. These Sealskins are unreal."

Colt Jackson assesses the situation around them. He realizes something evil is amidst. "Quick! Turn off your infrared!! Turn off your infrared!! It won't let you see the shark very well. The shark will be the same temperature as the water. It is THEIR CAMOUFLAGE!"

Petty Officer Paul Westwood, call sign Ocean Dragon, is looking around after turning his night vision off. "HOLY SHIT! Look at all the sharks!! There is black stuff coming into the water, also, from the sides of the lagoon. What's going on...??"

Stingray Charlie, adds, "There must be three dozen sharks I've counted already!! CRAP!! We only have 4 bang sticks and those will blow our positions!!!"

Papa Bear calms his men, "Be calm, gentlemen. Listen up. We unfortunately, sprung a booby trap---I think. The netting probably had a tension release. When we cut some of the filaments, it probably sprung some defensive mechanism---without electronics...those devious PRICKS!"

Ocean Dragon asks Papa Bear, "What's the black stuff, Papa Bear?"

"My best bet is blood. Lots and lots of blood!"

Petty Officer Duke Moran, Stingray Charlie, starts getting excited, "SHIT! Not good, not good. What are your orders, sir."

Lori Sanders watches the scene with her support staff in the NSA briefing center in Bahrain on the U.S. Navy Base. She unconsciously lifts her hand up to her mouth, in shock, "HUUHHH!" She says to herself, "*Lord, I hope Colt bandaged up that wound so blood isn't leaking into the water*!"

Inside the USS Minnesota in the radio room, Ghostship Commander Brian Tenclaw has taken the mike, "This is Neptune, Papa Bear, what is the sit-rep?

"Neptune, the lagoon has been loaded with sharks. They must have been collecting them from all over the Gulf, because I've never seen this many in one place in this area, before. The blood is putting them into a feeding frenzy. They are dashing about and the big sharks are attacking the little sharks!!"

"Papa Bear, this is Polar Bear. What do you advise?"

"Sir, the mission is compromised if we can't monitor the boat dock and beach. But, I have an idea."

Admiral Steven Whistler, call sign Polar Bear, asks, "What is it, Papa Bear?"

"Well, sir, when Neptune showed me around, very briefly, when we arrived on Ghostship, I saw the galley. Those boys eat pretty well---I think the guys were preparing steak."

Admiral Steven Whistler asks Colt Jackson, "What do you want to do with the steak?"

"Well sir, it's not the steak exactly, but if you could speak to Blue Angel in Bahrain, I'll bet he has a side, or two, of beef that could be helo'd in on the stealth choppers and dropped off around the bend of the nets---out of direct line of site from the beach at the hotel. We can try and evade here and wait for the chopper. It's only a 70 mile ride, as the crow flies, from Bahrain.

"Always thinking, Papa Bear! That's why I hired you!"

"Thank you, sir, but we have to get out of here right now. We're 500 feet from the net and our escape. It's going to be a long swim."

The Admiral is very concerned, "You better get going. The sharks will only get more crazed as blood pours into the water."

"When we get the beef, we'll set a trail up for the sharks outside of the lagoon and get them away. Once they taste filet mignon, or pork chops, they will target the source of the food. That will be our chance to get set up. We can cut a larger hole in the net to allow them to more easily swim out. Sir, the Persian Gulf, and especially lagoons, do not have many sharks. Frankly, we were taken by surprise. Sorry, sir. Can you have Blue

Angel send us some Shark Shields, also? That will electronically keep them away from us."

Sharks are sensitive to low-frequency electrical fields. When submersed, Shark Shield's electrodes create a three-dimensional electronic ring around a diver, using low-drag antennae which attaches to the SEAL. The strong electric current causes a shark's muscles to go into a spasm---especially around its nose. A shark's snout has electrical receptors called Ampullae of Lorenzini and they are extremely sensitive. Sharks use them to detect prey in the water at close distances. When the electrical impulse hits them, the shark is irritated and does not want to be near the swimmer. The SEALs simply strap the device to their ankles and are ready good to go.

"Will do, Papa Bear. I'm on it. No one could have predicted dozens of sharks in that lagoon, soldier. Don't let it get to you. Over and out." The Admiral immediately phones Vice Admiral John Stiller, Commander of the 5th Fleet in Bahrain, call sign, "Blue Angel." They speak briefly.

"Thanks, John. Let's see if we can get our Peking Gourmet friend out of this jam." Both the men hang up.

Colt looks at his digital clock on his Seaglass---11:37PM. Colt puts the SEALs in a ring formation, with their backs to each other and their bang sticks held out. They swim as calmly as they can to the hole in the netting.

"Easy boys, keep our strokes synchronized and gently. They'll sense the vibration in the water, for sure. Get ready with the bang sticks."

"Piece of cake, Papa Bear. Killer Whales eat sharks for breakfast!" says the Master Chief.

"Enough. Get serious. These things are out of control. Once they bump into something---they'll attack!"

Petty Officer Paul Westwood, immediately responds in a dead-serious voice, "Yes, sir!"

The SEALs swim about half the distance to the netting, about 250 feet, when disaster strikes. A massive Tiger shark has struck Ocean Dragon, Petty Officer Paul Westwood, seemingly coming out of nowhere, because his bang stick didn't go off. Ocean Dragon lets out a guttural cry, just like Killer Whale did.

"UGHHHHH!"

Ocean Dragon's bang stick falls to the lagoon floor and the entire SEAL Team is unlocked from its ring formation because of the force of impact from the Tiger. The monstrous shark, after hammering Ocean Dragon in the gut, swims by the SEALS at full speed, seeing nothing in front of it.

Ocean Dragon can't breathe. He is panicking and flailing in the water. All of his training didn't prepare him to come face-to-face with a 20-foot monster from the deep. He struggles for air and is making all kinds of noises, trying to regain his breath.

"P-A-P-A B-E-A-R-S-T-I-N-G-R-A-Y-I-C-A-N-T-B-R-E-A-T-H-E!"

He gags for air.

H-U-H-H-H-H!!

The Tiger is a behemoth. It makes a 180° turn, primordially remembering exactly where the soft flesh was it had bumped into. The shark feels the rapid movement of an injured prey. It accelerates through the water. The big Tiger has spent too many days in the lagoon without a meal, other than small sharks. With its four-inch, razor-sharp teeth opening wide, the huge beast strikes Ocean Dragon with all its might. The force of the impact grabs Ocean Dragon like a rag doll and the SEALs record, live on their Seaglass, as the mighty shark bites Petty Officer Paul Westwood's right knee. The cameras continue recording. The men in the water and everyone witnessing the events over the live camera feed are helpless to do anything.

Blood gushes everywhere! The cameras follow Ocean Dragon and the rear view of the shark, as the Petty Officer is dragged off into the depths of the lagoon.

Colt Jackson's SEALs watch in anguish as the figure of their friend and fellow-warrior can be seen, through the moonlit lagoon and blood-clouded darkness, being torn apart by nearly two dozen sharks. The Team knows this man just gave up his life for them. It is their chance to escape!

"Move it, men! Double-time. There is nothing we can do for Ocean Dragon right now. We have to evade and escape immediately! Move it! We'll be back and we'll hunt down that bastard Tiger shark when we salt the trail leading to the beef. We'll sit on the ocean floor and I don't care if we attack it with knives, we're going to kill that SOB!"

"Roger that," Killer Whale grimly states.

"He's the first friend I've lost in the SEALs, sir. We spent hours together and worked on all the technology. I can't believe he's gone."

"Snap out of it, Stingray! We've got a mission to complete!! That's an order!!" Colt is nearly shouting through the underwater COMM system.

"Sorry sir. I'm back."

Simultaneously Lori Sanders and her crew of technology wizards gasp in horror at the terrifying killing of Petty Officer Paul Westwood. The video from the other SEALs' cameras captured the behemoth vividly. Lori queried "Tiger sharks" when Ocean Dragon was struck. The NSA computers quickly analyze the images of this beast and Lori's screen pops back a measurement:

"Tiger.Shark.Image.Processed = 20 feet 6 inches."

Lori Sanders and her crew look at the screen readout and cry out in shock,

"HUUHHH!"

Everyone is clasping their hands over their mouths.

Lori types in "Tiger shark Record". The computer spits back "1,785 pounds and 20 feet long."

"Polar Bear, do you read me, this is Dolphin."

"Go ahead, Dolphin."

"Sir, the shark that just ate Ocean Dragon, sir, the computer indicates it weighed in at 1,800 pounds! The world record is 1785 pounds. The computer also indicated it was twenty feet, six inches long---also a world record."

"They must have penned these animals for quite some time. I better let the team know what they are up against, over and out." The Admiral switches to the secure COMM line to the SEAL unit in the lagoon.

"Papa Bear---the report from Dolphin was the Tiger is over 20 feet in length and weighs in at 1,800 pounds, by the computer's estimate. You don't know what else is lurking in those dark waters, get out of there!"

Colt checks his digital clock, again, on his Seaglass, it's 11:40PM in Qatar. He says to the Admiral, "We're nearly there, sir. They swarmed all over Ocean Dragon, sir. He gave us a way to escape. We're through the net now."

Lori Sanders speaks out loud to her group. "THANK GOD!"

The Admiral says, "Good work getting out of that. What is the mission status? Abort?"

"No sir, we still have three people to cover the main areas of the beach, the boat dock, as well as the angle from the jetty. I will get the remains of our fallen soldier and return them with us when the mission is over, I promise."

The Admiral tells the SEALs, "He'll be given a burial with the highest honors."

Colt replies, "He was a popular kid and very bright. I think there will be plenty of Budweisers on his coffin. We're through the hole, sir. How long before the helo drop? We're heading to the anchor point. We'll be there in five minutes."

"Budweisers" are the Navy SEAL nickname for the gold tridents they earn when they complete Special Warfare Training BUD/S and SQT, which stand for "Basic Underwater Demolition/SEAL" and "SEAL Qualification Training," respectively. When one of their brothers falls, he is honored with the pounding of his fellow SEALs' Tridents into the top of his wood casket. It takes up to an hour for all the SEALs on a Team to give their last salute to the fallen hero. Each man drives his trident into the wood. When they are done, a simple coffin is transformed into a gold-plated tribute to a fallen friend and brother-at-arms. The tridents are left on the coffin, reminding everybody who sees the sight---the SEALs are one.

At 11:51PM, Colt Jackson, Petty Officer Duke Moran and Master Chief

Jack Flintlock turn on their Pegasus Thrusters and swim to the Seahorse. They switch to its breathing system, preserving their MK 25 rebreathers.

Vice Admiral John Stiller, call sign "Blue Angel," sent out the Bahrain base alert team and they raided the storage freezer. "Blue Angel" is communicating through the Pentagon with Commander Colt Jackson.

"Papa Bear, Blue Angel here. Have two sides of beef and two pork bellies heading your way on a Stealth Black Hawk. They will be ready to take off in another ten minutes, or so. We also raided the blood bank. Use your Infrared Glo-sticks to bring the chopper into position. They'll lower the meat and then spread the blood in a line toward the entrance of the lagoon. I'm sorry to hear about Petty Officer Westwood."

"He was a very good man, sir. He would want us to finish the mission, that's for sure. He gave his life for us. By the way, thanks for the assist here. Looks like I owe you a much better dinner than the Peking Gourmet, sir. That's twice this week---you FEEDING ME on YOUR TAB. Sir, if you can order the chopper to drop the meat with a splash, just before it is ready to return to base, the noise and vibration will help attract the sharks."

"Will do, Commander. We'll spread the blood first from the net out to the splash point. Have your team light a few spots up with the Glo-stick Chemlights."

"Roger that, sir. If you can let the crew on the chopper know to drop the Shark Shields near the netting, we'll be there to retrieve them. Thanks, again."

"Good luck, Commander. Over and out."

Colt and his team sit and rest twenty minutes on the Seahorse, still hidden underneath the surface of the Persian Gulf and the eastern entrance to The Pearl. They switch their air back to the MK 25 rebreathers and switch on their Pegasus Thrusters. They arrive back at the entrance to the lagoon, in front of the net, in slightly more than 5 minutes. It is 12:15AM. The stealthy Black Hawks should arrive at about 12:30AM, assuming they are flying at their top-rated air speed of 196MPH. With an "as the crow flies" distance of only 87.41 miles, the actual in-air flight time from the U.S. Navy Air Base in Al Manama, Bahrain is only 26 minutes and 45 seconds.

Colt again reads his digital clock and knows it is now the next day---August 16, at 12:18AM at The Pearl. Colt and his team light up their waterproof Chemlights.

Commander Jackson orders, "Stingray Charlie, take four Chemlights and hustle out on a straight bearing to the Seahorse. Light up a Glo-stick every 400 hundred feet and let it float on the water in the middle of the channel. Start the first one about twenty yards from the net. There isn't too much of a current tonight, so they won't go anywhere. We'll spot our location here in an 'X' with the lights. The chopper will spot their infrared glow and drop the Shark Shields right on top of us."

The Chief COMMs Commander Jackson, "The old "X marks the spot, huh, Papa Bear?"

Colt replies to both men, "You got that right Killer Whale. The Black Hawk knows to drop the Shark Shields on 'X marks the spot'---I COMM'd them from the Seahorse. They will spread the blood along the lit path toward the fourth light---and I asked them to make some smaller splashes, with carved off meat along the way, as they approach the drop point."

The Chief squawks and looks through his Seaglass at Petty Officer Moran, "It shouldn't take you very long with the Pegasus. I think you can lay them down and be back in about ten minutes. Good luck, Stingray."

"I'll be back in 11minutes and 36 seconds, Chief! I already calculated it in my head. I'll lay out the lights on the way out and then hustle back."

Papa Bear instructs Petty Officer Moran, "The Black Hawks will be here at just past 12:29AM, so make it snappy. We don't have any time to waste. If there is a problem and you have to abort, or don't get the Chemlights all down, head to the Seahorse."

Stingray Charlie responds "Aye, aye, sir! I'll be fast---don't worry about *that* Papa Bear. The last thing I want is to be caught in the shark's feeding frenzy, again---especially with the Black Hawk creating a trail of bread crumbs for them to follow!"

Stingray Charlie takes four of the lights and swims off. He places the Glo-sticks along a path in the channel, ending about 1,000 feet from the lagoon. Using his Pegasus Thruster, he finishes this task in less than five minutes and speeds back toward the net. He keeps thinking in the dark water,

"I sure don't want to be along this path when that meat splashes into the water!"
At 12:29:30AM, the Black Hawk arrives and silently begins lowering the Shark Shields down to the infrared beacon of the lit "X" below. Papa Bear and Killer Whale swim over to the retrieval point for the shields. The Black Hawk is a welcome site to the SEALs.

Stingray Charlie has still not returned.

Colt Jackson waives a lit infrared stick signaling the Black Hawk where they are in the water. The chopper maneuvers directly over the SEALs, below.

"Stealth Raider, this is Papa Bear, we are directly underneath you, lower away."

The helicopter replies, "Aye, aye, Papa Bear, Stealth Raider out." The shark shields are speedily lowered to Papa Bear and Killer Whale.

"I've got them!" Killer Whale grabs the Shark Shield bag and releases the line. The Black Hawk hovers in near silence about 40 yards away and begins pouring out blood and throwing one-pound steaks out the open door of the chopper. They splash into the water below.

Stealth Raider creates a brilliant bread-crumb trail for the sharks.

"Stingray Charlie, you almost here? The meat is splashing, son, you better kick your fins and GIDDYUP! Keep the Pegasus on full power!"

"I read you, sir. I'm about fifty feet away and I see gray figures in the water, I'm veering off to your signal."

Colt Jackson has moved himself and the Chief away from the entrance the SEALs cut into the net, positioning them near the point of the Fanar Fayrouz, where some of the finest villas at The Pearl, and world, are located.

"I'll be there in a second, or two."

"Welcome back, Stingray," Killer Whale says. Upon arrival the Chief hands Petty Officer Duke Moran his Shark Shield. "Good job, Stingray! Here put this on, we're protected now. We'll give the sharks about 15 minutes to clear out and then we'll go back in the Lagoon and set up for surveillance."

Colt Jackson tells his men, "We haven't seen, or heard, any boats coming in, so we haven't missed anything from the Gulf. I'd say we're still on schedule, albeit 45 minutes late and missing one brave SEAL. Let's get moving and make this count for something. Stay alert---*because* it is the middle of the night, I expect guests to arrive. It gives them better cover."

"Roger that, sir," says Stingray Charlie.

"Aye, Papa Bear," says Killer Whale.

By 12:45AM, the lagoon is clear of sharks. The SEALs have watched a veritable parade of hungry sharks beeline out of the now larger hole in the net, which the SEALs had previously widened. It was a good thing they did. Petty Officer Duke Moran counted forty-nine sharks leaving the lagoon! About one-fourth of them were at least 15 feet, or larger.

Stingray Charlie makes the count, "...43, 44, 45, 46, 47, 48...49! Look at the size of those things! Did you see the Tiger go by?"

Commander Jackson says, "I sure did, Stingray. When we finish our mission for the evening, we'll keep an eye out and see if we can spot him around where the meat was dropped. Back at the Seahorse in the storage locker, I have two spear guns with cyanide tips, courtesy of Ghostship. I'd love to get some payback, but first things first. Let's focus on our mission."

CHAPTER 8

THE MARSA MALAZ HOTEL

With the sharks out of the way, the disciplined SEALs quickly go about their business. They have to set the memory of Petty Officer Paul Westwood aside for the moment and go about the task of implementing their surveillance plan. They all know he would want them to complete the mission.

Chief Petty Officer Jack Flintlock will set up surveillance in the boat dock area, and Petty Officer Duke Moran will set up surveillance on the Jetty, which protrudes from the beach of the Marsa Malaz Hotel. Colt Jackson will initially set up on the opposite side of the beach, on dry land. A number of cabanas and ornate, beautifully laid stone walkways, lead to a number of good spots to find cover.

The one area Colt wants to focus on is the courtyard, just up from the beach. If he can surveil this area, his team will have all the key areas covered.

With the Sealskins working beautifully, the SEALS find they blend into their respective positions in complete stealth mode. They are invisible.

"Aye to that, Papa Bear. I'm going to set up shop in the boat dock," says Chief Petty Officer Jack Flintlock, call sign Killer Whale.

"I've got the jetty," says Stingray Charlie.

"I'm heading for the opposite side of the beach from you, Stingray Charlie. I'll beach and surveil for roaming guards. Remember do not engage, unless we are engaged upon."

Both SEALs say, "Aye, aye, sir."

At the lagoon beach and courtyard of the Hotel Marsa Malaz, it is now 3:25AM on the 16th of August. During the last couple of hours, Colt has located active patrols all around the hotel. He has counted thirty-six armed guards patrolling the grounds with Uzi submachine guns. They are protecting the hotel with the utmost care. The guards look under chairs, check and recheck cabanas, and scour the island for intruders.

They check the perimeter with a special team whose sole mission is to concentrate on the large seawall built to protect the island from corrosion. The boulders which make up the sea wall, would provide an easy way for an intruder to secretly penetrate the island. There are eighteen men protecting the perimeter, alone. If the SEALs didn't have their Sealskins, they would have been discovered nearly immediately---no matter how careful they might be.

Colt Jackson makes his way into the courtyard area. There are beautiful palm trees all over the patio, with incredible landscaping around them. Flowers, water and pools create a tapestry of peaceful, elegant beauty in this area of the hotel. The view from the hotel entrance across the courtyard and into the lagoon is spectacular. Colt has worked his way right next to the hotel door entrance and is up against the wall. He is just to the left of the large, expansive and stately glass entryway. Suddenly, the large, glass door swings open and a guard walks out. Colt lowers his camouflage face covering and holds his breath. He flattens himself against the wall. The guard walks right past him on the walkway and doesn't see, or hear him.

The guard is smoking a cigarette and it brightly glows, lit up by the night vision capability of the Seaglass headgear. The GUARD walks right by Colt Jackson with his Uzi submachine gun pointed down, toward the elegant walkway.

"*There is no threat to the hotel at this hour of the morning,*" he thinks.

The Islamic State guard continues thinking to himself, "*Allah has truly blessed me this evening. It is quiet and has cooled off beautifully. Let us pray the arrivals go well at 3:30...*" Two feet past Colt Jackson, the guard all of a sudden, sets his Uzi on the ground and kneels. He lies prostrate, facing east. He prays a quick prayer of thanks to Allah. He hums a prayer in near silence,

"ALLAHU AKBAR…"

Colt Jackson thinks, "*Thank God I have this Sealskin on.*"

Back at the Bahrain COMM Center, Lori Sanders and her crew are watching the scene. Her pulse quickens when she sees the guard walk past Colt Jackson with the cigarette lit up brightly by the infrared. When the guard spontaneously gets down in the prone position, she nearly has a heart attack. Sealskins do not conceal noise and any little breath, or movement by Colt Jackson, will easily give away his position!

Lori is staring at the prone, praying guard. "You've got to be kidding me! Of all the places to stop and pray. I hope Colt can still hold his breath a long time."

Admiral Whistler is witnessing the same scene. He is staring at the prone, praying guard. He speaks to others witnessing the scene, almost unconsciously. "That boy sure knows how to poke his nose in a hornet's nest! Papa Bear always seems to know where to find trouble!"

The prayer has been going on for a couple of minutes, but they seem like hours to Colt. He can't hold his breath anymore and he expels it---a little too loudly. The soft, strange noise was noticed by the GUARD. He quickly ends his prayer, grabs his Uzi and stands up!

Lori Sanders and her crew of analysts are riveted on the screen. "Oh no, here we go again! Watch out, Colt!!" She says rather loudly, even though she knows Colt cannot hear her.

Back at the courtyard to the hotel, Colt's adrenaline starts pumping wildly. He fingers his trusty DOC Flipper Knife, which has saved him and others many times before. Colt is about ready to pounce on the Guard.

The guard pulls his Uzi up and scans the area around him. He still can't see anything, but swears he heard something. He thinks, "*It sounded like a breath.*" He speaks to the night air, while waving his gun around,

ما اسمك؟هو ما اسمك؟هو هناك؟ من

MONDA HUN DA KA? MAA ISMUK? MAA ISMUK?!!

(Who is there? What is your name? What is your name?!!).

Lori Sanders computers immediately translate the guard's words. "What is your name?" Instantly shows up on her translation screen.

The NSA can instantaneously translate any verbal, or written language, into English, or any other language at the speed of light.

On the secure COMM line to Admiral Steven Whistler, Lori says, "Admiral? He is asking, 'Who is there? What is your name? What is your name?' I think Colt may have his cover blown."

"I'll relay it on. Thank you, Dolphin."

The Admiral delivers the message to COLT JACKON---whispering into the encrypted COMM line to avoid any feedback of any kind at Colt's end.

"You better get ready, Papa Bear.
He's saying, 'What is your name?'"

Colt Jackson does not reply, but stands deathly still.

The guard again says,

"MAA ISMUK?? MAA ISMUK???"

The guard is turning in a 360 degree circle. He still can't see anything. The Uzi's rifle muzzle comes within an inch of Colt Jackson's Sealskin.

Suddenly, the roar of a powerful, sleek yacht hydrofoil comes into the lagoon, startling the guard. The Islamic Jihadist menacingly turns in another 360-degree circle. He sees nothing, and then sprints down to the beach. He cannot be late for his beloved leader's arrival! Colt Jackson takes a very deep breath and moves his body around to loosen it up. He was starting to cramp up.

The Islamic State guard is talking to himself, as he runs toward the sound in the lagoon, "Maybe it was the wind. I have to get to the beach and get the guard in order. The Caliph is arriving!"

The Admiral says, "That was a close call, Papa Bear."

"Yes, sir. I'm going to kiss whoever invented this suit---I don't care if it was a man, or a woman! It saved my butt!!"

"I'm sure they would really enjoy that, Papa Bear!" The Admiral says sarcastically. He and Colt enjoy a good laugh.

"Very funny sir. I'll at least give him a huge hug!"

Lori is busy on her computer. She just pulled up the classified patent for the Sealskins. It lists a number of scientists from Canada as the inventors.

The first person is a woman. Her name is Yvette Lemieux. Lori turns to her team. "Get me everything you have on this person. These suits are amazing. We need to pull these people closer to us and put them permanently into our orbit."

Kristi Kruger responds immediately, "I'm on it, Lori. I still can't believe I watched the guard come within an inch of Commander Jackson and not even see him! Truly remarkable. I'll have the report to you within an hour."

"Thank you, Kristi."

At 3:40AM the Emir's sleek hydrofoil and its powerful engines come to an abrupt stop in the lagoon. Simultaneously, out of the boathouse, roar eight Iguana 29 Luxury Superyacht Tenders. Their carbon fiber bodies, made with advanced resin infusion technologies and powerful German engines, give them 300 Horse Power and allow speeds of up to nearly 50 MPH over the water.

Colt says to the Chief, "We have action on the hydrofoil."

"Roger that sir. Eight drivers just came into the boat dock and took off in eight Iguana craft. It looks like they're heading to the hydrofoil. I captured them on the cam. Over."

Unique to the Iguana, it is amphibious and also extremely luxurious. It is built for billionaires. But, these particular eight Iguanas were built for the Emir. They came with solid-gold stairways, cleats, and rails. The table on their decks was designed by Vivien Durisotti, one of the leading designers in Paris. Crocodile leather adorns the seats, storage drawers and even the wheel and controls. Bose, water-proof surround-sound, rings the passenger area.

Iguana 29 Exclusive

Photos courtesy of Iguana Yachts. http://www.iguana-yachts.com/

Iguana 29 Exclusive

Photo courtesy of Iguana Yachts. http://www.iguana-yachts.com/

Hidden in the hull of the Iguanas are tracks---similar to those used on a tank, except the rollers are made from Kevlar with reinforced, nylon-filled carbon, rubber. The axles, lifting mechanisms and sprockets are all made from stainless steel and the lifting arms are made from marine-grade 5083 aluminum. The tracks are made to operate on many kinds of ground, including the soft, shifting terrain of sand and marshes.

Because of the many islands and shallow water in the Persian Gulf and Middle East, these craft are ideally suited for landings in shallow water. The Hotel Marsa Malaz does not have a dock directly on its beach and the Iguanas make the perfect delivery vehicles for the mystery guests.

The guards collect together, and stand at attention before the lagoon. The Iguanas approach. The roar of their eight engines is deafening. As they roll up onto the sand, their tracks appear out of the water. When they return to the water, their tracks retract in only a few seconds. This converts the Iguanas back into very fast boats, with perfectly flat hulls---ideal for the shallow waters of the islands in the Persian Gulf.

Eight figures are clearly visible in the moonlight on the hydrofoil. They are picked up---one person per Iguana which, in and of itself, was hugely conspicuous. The SEALs observe this spectacle and knows each Iguana can easily hold eight passengers. Whoever is coming to the hotel at this hour of the morning is clearly important.

Colt Jackson gets into position along the entryway to the courtyard. He will be in a perfect place to catch the faces of the passengers, as they walk into the Hotel Marsa Malaz.

On the beach-front entrance to Marsa Malaz Courtyard at 3:45AM, Colt Jackson presses up against the courtyard wall which joins the beach entrance to the hotel. He will focus his Seaglass on capturing the visitors who come off the Iguanas.

"OK, Killer Whale…" Colt turns on his zoom in his Seaglass to look at each Iguana riding up to the beach. "…contact the inbound SEAL Team and let them know, we will be running about 30 minutes late. On my order, have them remain at the anchor point."

Chief Petty Officer Jack Flintlock squawks, "Starfish 2, this is Starfish 1, over." Starfish 1 is Colt Jackson's surveillance unit, Starfish 2 is the next team to take over the shift.

"Starfish 1, we read you, over."

Starfish 1 replies, "Is that you, Tiger Fox?"

"Affirmative, Killer Whale, we drew the dawn shift. How you doing? We're all sorry to hear about Ocean Dragon."

Killer Whale stays on his business, "Papa Bear wants you to hold position at anchor point for another 30 minutes. We just got hot here."

Lieutenant Cody Custer replies, "Roger that."

Killer Whale asks, "Do you see any sharks around?"

"No, nothing, but they are still a bit away from us. After losing Ocean Dragon, all of us grabbed Shark Shields from the sub. They are all turned on. It sounds like it was pretty nasty."

"You've got that right. An enormous hammerhead originally bumped into me...I could barely breathe. It was absolutely frightening to watch how quickly the Tiger regained its composure after striking Ocean Dragon. With his Sealskin's active camouflage turned on, the Tiger Shark was startled after it struck. But, the shark then did a 180-degree turn and slammed right back into Ocean Dragon at full speed---with its jaws wide open. The Tiger knew exactly where his prey was. Ocean Dragon didn't have a chance, the poor kid. I never want to witness another shark attack up close and personal like that again, EVER!"

"I hear you, Chief. He was a great kid and super sharp. A Budweiser honor guard for him, that is for sure."

"Roger that. We'll see you in 30, unless something else happens. Papa Bear wants you to standby until ordered otherwise, or relieved."

"Aye, aye. Over and out."

At the beach-front entrance to the Marsa Malaz, Colt trains his Seaglass on the inbound Iguanas. They are going to come in, one after the other, in line with their departure from the Emir's luxurious hydrofoil. It should be perfect for the Seaglass. Colt watches as the first Iguana comes up out of the water. Its Kevlar-strength tracks magically appear, now protruding out of the hull and grabbing at the sand. The Iguana effortlessly climbs up the beach onto dry land. Colt zooms in his Seaglass on the pilot of the Iguana, and the sole passenger in the first craft to land.

Exclusive Iguana 29

Photo courtesy of Iguana Yachts.
http://www.iguana-yachts.com/

"Polar Bear, are you getting this, sir? We have the Caliph, himself, stepping down the steps of the Iguana. He'll walk right by me in my position near the wall."

Admiral Steven Whistler, from the Pentagon, replies, "We see with picture-perfect clarity. The lights on the beach are bright. You should get a great shot of him."

"Yes, sir, I didn't need the night vision on, so I turned it off. The beach just lit up when the Iguana's began leaving the mother ship. They obviously don't want their guests to trip over a lounge chair in the darkness!"

Colt continues capturing the guests via HD video. The NSA-linked facial recognition software works flawlessly. The first to arrive is Caliph Abu-Bakr, one of the most dangerous men on the planet, who singlehandedly united the Islamic State. When he steps down from the boat, every guard cries out in exclamation.

ALLAHU AKBAR!; ALLAHU AKBAR!; LA ILLAHA IL ALLAH, ALLAHU AKBAR!; ALLAHU AKBAR, WALILAHIL-HAMD!

Colt's Seaglass translator is now turned on. He had forgotten to do this in the middle of the night when the guard was on silent patrol. It is now translating, automatically, the Arabic on his display.

> "Allah is Most Great; Allah is Most Great; there is no god except Allah, Allah is Most Great, Allah is Most Great, and all praise be to Allah."

The chief Islamic guard states, "Oh, Great Caliph! What an honor Allah brings us this fine morning! Welcome, your greatness!"

The Caliph replies, "Allahu akbar! Thank you for your warm greetings. As-salamu alaykum."

When the Caliph speaks, Colt's Seaglass screen translates the Arabic,

> "Allah is Most Great! Peace be with you!"

As the important visitors come into focus with the Seaglass, names flash up on Colt's screen..."General Dmitry Zhirov, Vice-Premier Wang Kai, General Kim Se Bong, Admiral Muhammad Mubarak, Sheikh Yusef Al-Masri, Abdul Qadeer Khan," and finally, on the last Iguana, out steps General Jose Lopez, the man working as a double-agent for the United States Government.

The eight men are greeted royally, by the Emir's personal staff. Some of the most beautiful women in the world are now lined up on both sides of the walkway up to the hotel. They are all virgins, brought in by the Emir of Qatar, himself, Sheikh Muhammad Al Therani, for his honored guests.

"Polar Bear…request permission to take them all out right here and now. It is a collection of the most vile, most radical Muslims on planet earth, sir. If we take them out, we can behead ISIL, immediately." Colt Jackson cannot resist the urge to try and end the war with these fanatics, right here and now."

"That is a negative, Papa Bear. We must learn what they are discussing. We must learn from General Lopez what is happening inside the hotel."

"Polar Bear, you have to admit these Sealskins work great, right? The autonomous camouflage, which no one really knew how well it would do when submersed in sea water for so long, works perfectly. I am invisible with this on."

"What are you proposing?"

Colt Jackson replies, "I just follow them in. When they all head to their rooms, I'll tag behind the General and find out when the meeting is going down and WHERE. Maybe we can figure out a way to get ears inside."

"Papa Bear, hold on. I'm going to confer with Four Eyes."

Inside NSA Headquarters in Ft. Meade, Maryland, General Shane Maxwell is observing the operation. He looks at his clock on his wall. It is now August 15th, 8:58PM Eastern Time, and August 16th, in Qatar at 03:58AM Qatar/Bahrain time. The General's office is filled with advanced wizardry of all kinds, with a complete video wall monitoring every single video camera. It is a crucial mission and he has been monitoring Starfish's progress. His secure line rings, he knows it is Admiral Steven Whistler, his friend and fellow officer.

"Hey, Steve, those Sealskins are something special. I say give Colt the AOK."

"That is my take, as well, General. Thanks for delivering them and getting them developed!"

"My pleasure, Admiral. If we can save peoples' lives, I am all in favor of it, especially these kinds of men, who risk their lives for everyone in the country. I'm still amazed at their courage. I just wish we could tell 330 million other people in America the kinds of risks these men take, just to keep our country free."

"Amen to that, Shane. I'll speak with you later. Thanks." The Admiral switches to the encrypted COMM line to Papa Bear. "Papa Bear, you are a go!"

"Aye, aye, sir. That's a good call by both of you, I'm sure of it."

The eight honored guests walk through the courtyard and into the luxurious halls and marbled elegance of the Emir's Marsa Malaz Hotel.

The guests are led by several guards and are also followed by several others. Colt Jackson slips into the line---directly behind the last guest, General Jose Lopez.

They all proceed into the hotel. No one has any baggage---it wasn't allowed on the hotel property for security reasons. No one needs to check in. Before they are all individually escorted to their rooms by the hotel's world-class hotel staff, everyone, including Caliph Abu-Bakr must walk through an interactive X-Ray and chem-sniffing machine. No electronics, no weapons, and no brief cases---no items, of any kind---are allowed.

As part of the Emir's and Caliph's security control, everyone meeting had to exchange their clothing for the luxurious thawbs the men are now wearing. These robes were provided by the Emir, himself. They are laced with gold and made from silk. They are very smooth and elegant to the touch.

All of the honored guests make it through security without problem. Colt Jackson follows General Jose Lopez and the beautiful staff member, to his room door. The General and the beautiful woman enter an enormous suite and she shows him around. They are speaking in English. The door is slowly closing.

Colt sticks his foot in, just before it closes shut and slips in without them noticing. The camouflage system is active and he blends into one of the alcoves, which is connected to the master bedroom. The woman is showing the General the view over the beach, below. It is now 4:20AM August 16th at The Pearl's Marsa Malaz Hotel in D'oha, Qatar.

She is talking with the General. "I am Fatima, General, anything you want---ANYTHING at all, you are to have. We are all also fighting the Jihad and play our roles. If we can give pleasure to our leaders, who risk our lives for us every day, we do so with the blessing of Allah. My body is your slave. So is any other woman, or any other thing you might desire here. If you want little boys, or little girls, the Emir has them also."

The General's Islamic heritage is not radical. He views these kinds of desires as pure evil, secretly. "Thank you, Fatima. Allah is truly blessing me and I will not fail to call you. Your beauty and refinement is a gift from heaven."

Fatima is blushing and smiling at the General. "Thank you, General. It would truly be my pleasure to make you feel at home and comfort you."

The General responds, "I've been awake for nearly 24 hours and I know we start the big meeting on the 17th at noon. Do you think I can get a rain check?"

Fatima is clearly saddened by this response, "I am not beautiful enough for you?"

"OH NO, Fatima. You are stunning and gorgeous. I am truly just exhausted. Caracas is a long, long way from D'oha and The Pearl. I must be fresh for tomorrow. It is of utmost importance to fulfill the will of Allah.

Fatima smiles with understanding. She likes the General. "I understand, General. Allah's mission is, of course, the most important. I await you beckoning me. Sleep tight. Remember, my calling is to be a comfort woman. I have preserved myself until now for our bravest leaders. I wait for you to summon me. I am a virgin."

The General gulps. "Fatima, PLEASE know I long to be with you. Please don't work too hard. If you can---please save yourself for ME. I would love for you to comfort me. Perhaps tomorrow after the meeting?

Fatima lights up like a light bulb. "It would be AN HONOR to hold you close to me. I will do anything and everything to you. Insha'Allah..."

"Insha'Allah, my beautiful Fatima."

Fatima exits the room and General Jose Lopez escorts her out. Before leaving, Fatima gives the General a hug. He feels her body rub against his and he gets excited. She shuts the door of the room behind her and walks down the hallway toward the elevator.

Colt Jackson is thinking facetiously to himself, "*So, this is how they recruit them!*"

Lori Sanders and Kristi Kruger stand with their mouths gaping wide open. One of the crack NSA software technology wizards in the room shouts out, sarcastically. He is known for calling them like he sees them, with a pretty good sense of humor to boot.

Brian Pierce shouts out, "Oh man, SIGN ME UP FOR THE JIHAD! If we had those kinds of perks, productivity would go SKY HIGH! Can I be excused, I've got a flight to catch!"

Brian is also based at the Glass House in Ft. Meade and works for Lori Sanders. He is a co-worker of Kristi Kruger. Both women see him grinning from ear to ear. They shoot him a dirty look and roll their eyeballs.

Kristi Kruger, who can dish it right back out with her quick wit, gorgeous smile, brown eyes and beautiful, wavy long brown hair, fires back, "DREAM ON, BRIAN! In your wet dreams!!"

Kristi hasn't figured out, yet, how to disguise her beauty. She is also an athlete and a classic beauty. Even with national security at stake, every day she turns men's heads in the Glass House.

She doesn't take any crap from anyone, however. She grew up with four brothers and has learned how to deal with aggressive cat calls, as well as overly-aggressive men. She is a Black Belt in Karate and held a 3.9 average on a 4.0 scale. She majored in mathematics at Yale and graduated three years ago. Her figure is one which most women envy.

Lori Sanders turns to both of them, "All right, you two. Stay on mission. Colt and his team may need our help. He has those new mosquito bugs with the advanced video and audio feeds. I thought he wanted them for his own team, but now I see what he's doing. He's going to give them to General Lopez and see if he can let them loose in the hotel."

Neither Kristi Kruger, or Brian Pierce, know anything of the extent, or nature, of any kind of personal relationship Lori Sanders has with Colt Jackson.

Kristi is just mesmerized with Colt. She says, "He's a real man, isn't he, Lori?"

With a stern look and raising an eyebrow at the young, 25-year old beauty, Lori says, "As I said, Kristi, STAY ON MISSION!"

"Yes ma'am. Sorry, seeing these guys risk their lives this way is just surreal. It really shows you what they are made of. I am staying on mission, Lori, but just between us girls, I hope to be with a Colt Jackson someday. What a dream!"

Lori Sanders feels a flash of heat come over her body and she nearly falls over. She puts her hand on her forehead and inhales deeply, then exhales. "Ho…ahhhh..."

Kristi innocently asks, "Are you OK?"

"I'm fine and honestly, I do know what you mean. Now, let's get back to work."

"You've got it, Lori."

Inside the General's suite, it is now 04:30AM in Qatar. Colt quickly sweeps the room for bugs and electronics with his Sealskin camouflage activated. He walks around like the invisible man inside the suite. The General heads to the bathroom and then washes his face, getting ready for bed. When Colt has finished sweeping the room, he finds no bugs.

It is now safe to turn off his camouflage and speak directly with the General. The General has his eyes closed and is cleansing his face. He has soap all over himself and is rinsing it off.

Colt whispers to the General, "General, it's me, Colt. Don't be alarmed. Remain calm and quiet."

Startled by Colt Jackson, the General lifts his head too quickly and bangs it slightly on the large, overhanging gold faucet. "Commander Colt Jackson??"

Colt says, "How are you, my friend?"

"YOU NEVER CEASE TO AMAZE ME! How did you get in here?"

"It's better you don't know, sir. But I have some presents for you. I think you will like them."

Colt holds up a mosquito-sized robot capable of taking HD video and audio recordings. The recordings can be stored locally, with wireless upload, when the mosquito is near a cell tower or in range of a network.

The General examines the nano-bot. "I'll be damned. A mosquito?"

Colt notices that while living in Venezuela, the General has readapted his Spanish birth accent.

"All you have to do is press the stinger and the mosquito's built-in GPS and radar will keep it afloat for days. It is programmed to leave this hotel on the 19th. We'll set up a homing beacon to recover them. They fly right in for a landing. It has already been programmed to find the outline of doors and outside sunlight. We'll be standing by to retrieve their transmissions when they exit the hotel. Once they have transmitted their data, they self-destruct. Pretty cool, huh?"

The General replies, "Our friends at the CIA and NSA are working together, nicely, I see."

"Yes, they are. But this little baby just came out of Operation PRISM. The people heading this up are beyond great."

Kristi Kruger shoots Lori Sanders a look. "Did he just pay you a compliment? I mean he KNOWS we're watching, right??!!"

Lori is blushing, but with a big smile on her face. She knows this is Colt's clever way of saying a direct "hello" to her. He knows she doesn't have mission access audio capability and unless critical, Lori stays off the SEAL COMM. It will only be used if authorized by Admiral Steven Whistler.

Lori, gathering her wits about her, after thinking fondly of Colt, says, "Sometimes in the heat of a mission, the people truly forget about the technology and just do their jobs. Colt is just doing his. He is building the confidence of the General. Colt knows the General will not take the risk of blowing his cover, unless he is certain the device will work."

Brian Pierce blurts out, "That's pretty good tradecraft, from what I can see."

Kristi says, "I get it. Yeah, it's pretty good thinking."

Lori continues, "Anyway, the effort to miniaturize those devices through nanotechnology took us years to develop. I led one of the key design teams which created the GPS and radar algorithms used to make the mosquitos intelligent. The navigation system integrates video, audio and GPS against radar mapping to plot an intelligent course."

Brian asks, "You mean it can differentiate between a person and a dog, or between specific people?"

Lori replies, "EXACTLY. For a mission like this, it uses a preloaded database of biometrics. When it is flying freely and has interactive network access, it can actually COMM directly with our computers here, providing global recognition search capability on the fly."

"That's incredible! You did that?" Kristi asks her boss.

"I lead a TEAM of brilliant engineers who created it. They did most of the heavy lifting, but I did solve one of the major stumbling blocks involved with integrating the audio and video data onto the device. I integrated a new encoding algorithm which compressed the data by a factor of a million to one, using nanotechnology and quantum physics. This solved the problem of locally storing large quantities of information, created by video and audio feeds."

Brian Pierce states the obvious, "IMPRESSIVE, Lori. I HAD NO IDEA. Regular computer compression of a file is only about 5:1, at best for large data files. That's an INCREDIBLE breakthrough. I hope you got a raise!"

Lori laughs. "Let's just say all of us work here for God and country."

Kristi says, "I didn't think we were able to mix God and government?"

Lori answers, "We all have religious freedom and can worship as we please. It is guaranteed by the Constitution. I always, personally believe, I am working for God and country. I am a committed Christian and proud of it."

Kristi Kruger likes her boss. "I know that, Lori. You are awesome. I mean, I knew you were brilliant, but I had NO IDEA I was working for a savant---as well as someone who is really nice!"

Brian Pierce snips at Kristi, "Brown noser!"

"I'm not brown nosing, Brian. You and I have both talked about it. We both really enjoy working for Lori."

Brian, somewhat reluctantly, but with a smile, says, "Well, OK, I guess I have to admit that one."

Lori responds to both of them in a nice, but firm tone. "GET BACK TO WORK AND FOCUS ON THE MISSION---PLEASE! But, thank you. You are both kind to say so."

Back in the General's hotel room, General Jose Lopez asks Colt, "How many do you have for me?"

Commander Jackson holds them in the palm of his hand. "There are six of these little babies. Just make sure you let two lose in the meeting room. The mosquitos' speakers and cameras are super sensitive. You won't even notice them. They will fly at ceiling level."

"Amazing. OK, I've got the picture, Colt. Now, how are you going to get out of here?"

"General, this is to keep you safe and it is for your own protection. I would like to ask you to go into your Master Suite for a couple of minutes while I make my way out. If I get caught, I don't want anyone seeing us together. I will be thought of as an infidel spy. You know what they do to 'infidels,' right?"

"I understand. OK, my friend. Shall we meet in Aruba, again, sometime soon? Those women were beautiful there!"

Colt, knowing Lori Sanders is listening to every word responds, "Yes, but too bad we didn't have any time to hang out. I'm sort of off the market, anyway."

The General is shocked. "YOU ARE?!! With who?"

"Well, let's just say it's a complicated situation." Colt thinks of both Lori Sanders and Mary Barrister.

In the meanwhile back in Bahrain, Kristi Kruger speaks out in shock. "HE'S OFF THE MARKET!! DID YOU HEAR THAT!! I'M HEARTBROKEN!!!"

Lori Sanders is now perspiring. "I'M SHOCKED, TOO. WHERE DOES HE EVER FIND THE TIME, I WONDER?" She smiles to herself.

Brian Pierce pipes in, "I want to ask him about the cute babes in Aruba. Do you think I can use the satellite this weekend---after the mission is over on the 19th? I'd love to check out the beaches there!"

Lori Sanders shoots Brian another dirty look, but knows Brian is kidding around. "BRIAN! KNOCK IT OFF!"

Brian Pierce starts laughing. "I can't help myself. I just love to hit both of your hot buttons sometimes! It makes me laugh, and in this job, we do need laughter."

Kristi responds somewhat sarcastically, "You're way too funny, Brian! Let's get back to work. We should have a treasure trove of information inbound three days from now."

At 04:40AM in Qatar, Colt Jackson shakes hands with General Jose Lopez. They say their goodbyes. Colt watches as the General disappears into his master bedroom suite. Colt walks to the door, takes one last look making sure the General isn't peaking. He then hits the switch which makes his camouflage active.

He vanishes and then opens the door to the suite. There is a maid standing by in the hallway, on call 24 hours. Each works an 8-hour shift. Every honored guest has a maid assigned to their suites. The maid watches curiously as she sees the door open. The General must want something.

She sits at her station, alert to serve him when he comes out the door.

The maid speaks to the opening door, "As-salam alaykom."

She sees no one. Colt hears her greeting and does not open the door all the way. He opens it just enough to squeeze through the opening and makes it appear the General opens the door slightly, then closes it quickly, changing his mind.

Colt walks down the thickly carpeted Persian rugs which adorn the hallway. The maid sees nothing and hears nothing. Colt heads to the elevator bank around the corner and departs for the lobby entrance.

At 04:45AM Colt appears at the beach-front entrance to Marsa Malaz Courtyard. He gets on his COMM to Lieutenant Cody Custer, "Tiger

Fox."

"Tiger Fox, this is Papa Bear. We're ready to be relieved. Mission accomplished with mosquitos."

At the anchor zone of the Seahorse, Lieutenant Cody Custer replies to his Commander, "Great work, sir! We'll be in the lagoon within ten minutes."

"We'll stand by until you are on site. Afterward, we'll head right out. Good luck Tiger Fox!"

"Thank you, sir! See you in a few minutes when we relieve you."

Colt Jackson, Chief Petty Officer Jack Flintlock and Petty Officer Duke Moran meet and greet their SEAL brothers underwater. The teams swim by each other, giving each other the "thumbs up" signal.

Kristi Kruger, viewing the scene says, "These guys are just way too cool. I've GOT to meet one of them."

Lori says, "If that's what you want, you'll meet some of them, one day. But, you've already met Colt…what are you talking about?"

Kristi Kruger, "Yeah, but HE'S TAKEN!"

Lori's face widens into a wide smile.

At the beach-front entrance to the Marsa Malaz courtyard, it is now two days later, on August 18 at 12:00PM Qatar/Bahrain time, 03:00AM Eastern Time.

Lieutenant Cody Custer is just finishing the Starfish 2 surveillance shift, on the third day of the Islamic State meetings at the Marsa Malaz Hotel. He has his active camouflage turned on and is sitting on a lounge chair, keeping watch on the hotel entryway to the courtyard. The rest of the Team is stationed at the boat dock, the jetty, and another SEAL is scouting the grounds.

Suddenly, the doors open and the entire group of Islamic State leaders emerge from the hotel, surrounded by beautiful women. They are all carrying beach towels for their men.

Fatima walks right by Tiger Fox, with her hand holding onto the arm of General Jose Lopez. Caliph Abu-Bakr leads the procession with the Emir. Sheikh Yusef Al-Masri, the moderate Cleric Chairman of the International Union for Muslim Scholars, based in Istanbul, follows right behind the Emir and the Caliph.

The beach, as it has during the entire time of the summit, has no less than six guards protecting the beachfront from any unwanted guests. The protection has proven useless, of course, against the U.S. Navy SEALs.

As the group approaches the guards, they surround Sheikh Yusef Al-Masri. He becomes frightened.

Sheikh Yusef Al-Masri cries out, "What is this! Take your hands off me!"

The guards strip off the Sheikh's thawb and underneath lays his fat belly and skimpy bathing suit. Both the Emir and the Caliph turn around and face him. The rest of the group forms a circle around Sheikh Yusef Al-Masri.

Caliph Abu-Bakr stares at him, looking seriously at the chief Islamic State guard. "Muhammad, hand me my scimitar."

The guard hands him the sword. "Here it is, my great Prophet."

Sheikh Yusef Al-Masri is pleading for his life. "Please, brother Caliph. Why? Why are you doing this to me? Have I not helped the cause by counseling moderation and unity? Our numbers have grown by millions!"

Caliph Abu-Bakr, with extreme contempt spits at the Sheikh, "You corrupt enemy of the true followers of Allah! You speak of moderation among the people of Allah? As ISIL was formed by the SUNNI people, the Islamic State is called to be a Sunni people. It is blasphemy to corrupt the will of the Prophet Muhammad and the will of Allah!"

Inside the USS Minnesota in the CSRR, Colt Jackson is awake in the radio room, along with Commander Brian Tenclaw.

Colt speaks into the mike, "Tiger Fox, we're catching this. This is great positioning. This Intel on the Sunnis is really very important. Don't respond, we can see how close you are to the scene. I just wanted to let you know, great work."

Lieutenant Cody Custer quietly taps his fingernail on his mike to let Papa Bear know he heard him.

Colt turns to everyone in the CSRR and speaks over the COMM line, for all to hear. "People, this is going to be gruesome. For those who would rather not look at the sight of a beheading with the arched blade of a scimitar, now is the time to get some coffee."

In the Bahrain NSA COMM center, Lori Sanders, Kristi Kruger and Brian Pierce are all watching the events unfold. Each of them feels their stomach start to turn.

Lori looks at Kristi Kruger. "Kristi, you want to get some coffee? Brian---are you OK?"

"I'm fine Lori. It will be gruesome, but I'll keep watch."

Kristi turns to Brian, "Thank you, Brian! I don't think I can bear to watch this." She looks gratefully at him because one of them needs to keep watch.

Brian has a "Clark Kent" moment of goofiness because he is smitten with Kristi. He manages a smile and wonders if Kristi looked affectionately at him. "Sure, Kristi, take as long as you need. I've got it covered."

By only 12:08PM, minutes after he was surrounded, Sheikh Yusef Al-Masri begs for mercy from the Emir Sheikh Muhammad Al Therani. "Please, great one. Have mercy!"

The Emir Sheikh Muhammad Al Therani looks at him furiously, "I cannot go against the will of Allah and our reincarnated Prophet and Caliph!" He looks with pride on Caliph Abu-Bakr. "Who do you think trained him in his youth? I DID!! By Allah's hand he led me to take in this brave warrior. He is now a leader of men. I TRAINED HIM IN THE TRUE FAITH. THE ONLY FAITH!! This is the faith of the SUNNI WARRIOR. We are sworn, before Allah, to be against all infidels who do not follow our beliefs and traditions. Allah will not bless us if we do not follow His will. THERE SHALL BE NO COMPROMISE!"

He bows his head to Caliph Abu-Bakr and steps aside.

Looking at Muhammad, the chief guard, the Caliph nods his head. The

Islamic guard binds the hands of the Sheikh with plastic handcuffs and forces him to kneel.

The Caliph orders, "NOW, ACCORDING TO ALLAH'S WILL---EVERYONE, TOGETHER!"

The Islamic State leaders, all the women and the Islamic State guards cry out in unison,

ALLAHU AKBAR!
ALLAHU AKBAR! ALLAHU AKBAR!

The Caliph continues, "While you Muslim clerics and scholars may claim Istanbul is the head of the Caliphate, there is only one whom Allah has chosen to lead it! ME! ALLAHU AKBAR!" He raises the scimitar's razor sharp, curved blade over the head of the Sheik, and swings down, with all his might, and shouts, again,

ALLAHU AKBAR!"

The Sheikh's head rolls in the sand and stops two feet from the feet of Lieutenant Cody Custer. Blood pours all over the beach.

Emir Sheikh Muhammad Al Therani speaks to the Islamic guards, "Throw him in the middle of the lagoon, the sharks will get him! Guards, REMOVE THE BLOOD! We need to celebrate the purging of compromise from our ranks."

The Caliph looks at the group of men, before him, "I know I do not need to warn any of you, my brothers. None of you has compromised on our faith. To be victorious against all infidels, we must not compromise. This is the will of Allah!"

The Islamic State leaders, all the women and the Islamic State guards in unison, in complete submission,

"ALLAHU AKBAR!"

The entire group of leaders and their women leave the beach and head for the luxurious indoor pool of the hotel. The Emir loves its inlaid silver and its imported Italian marble, with solid gold trim. The pool is one of the Emir's favorite delights. While the leaders approach the entrance to the hotel from the courtyard, Lieutenant Cody Custer sits deathly still in his lounge chair, while one of the guard comes over and grabs the cut-off head of Sheikh Yusef Al-Masri.

Two other guards grab the Sheikh's body and they walk to the far side of the beach. Along the jetty, they dump the body into the lagoon. The entire scene is captured by the Starfish 2 team. As the guards walk toward the jetty, Lieutenant Cody Custer invisibly slips in behind the Muslim leaders, and their entourage, to record the scene at the pool.

The indoor pool of the Marsa Malaz Hotel sparkles. The leaders are happy to be rid of the compromising cleric Sheikh and celebrate his beheading with the bountiful things of beauty and elegance the Emir has arranged for them.

Vice-Premier Wang Kai tells the Caliph, "We are wise not to compromise. Thank you and thanks to the Emir, for alerting us to your plan to rid ourselves of this corruption. Disarming the cleric, by pretending to go for a swim on the beach, was a stroke of genius prior to his beheading."

"Thank you, Premier Wang Kai."

The Emir responds to the Caliph, "Thank you, great one. It is an honor to protect our faith. Now, let us enjoy our company and one of the finest pools in the entire Middle East. MUSIC PLEASE! DRINKS! Now, we go into the pool and celebrate our greatness, together. It is our last day before our destiny is fulfilled. Eat, drink and be merry! Allahu Akbar!"

Everyone cheers and all the men grab multiple women. General Jose Lopez has taken Fatima. He is very taken with her.

From his office, Admiral Steven Whistler observes the goings on with interest. He is especially observant of General Jose Lopez. The women strip naked and an orgy appears to be taking place in the pool, with men and women intertwined with each other.

On the Ghostship, USS Minnesota, after viewing the beheading on the sub's holographic video screen in the COMM room, Colt Jackson orders his unit, Starfish 1, to grab some down time and shut eye by 6PM. They will be up in the middle of the night to complete their mission the following morning. He wants his men to relax. Beheadings can make even the bravest warriors squeamish. Colt wants them to put their mind on other things at the moment.

On August 19th, dawn is at 05:08AM in D'oha, Qatar. It is two hours before, nearly 03:00AM. Colt Jackson's unit, Starfish 1 shift, is taking the day-break surveillance shift on the last day of the mission. The mosquitos are extremely sensitive and Colt wants to make sure they are not only handled properly, but also are uploading the data immediately back to NSA in Bahrain. Whatever has been discussed the last few days is likely to be highly important for national security. The mosquitos are preprogrammed to depart the hotel on the 19th at 10:00AM. Colt and his men are rechecking all their gear and testing their electronics, prior to leaving Ghostship for this last part of the mission.

Colt, Chief Jack Flintlock and Petty Officer Duke Moran swim out of the dive chamber to the Seahorse for one last excursion to the lagoon and anchor zone. By now, everything is working like clockwork.

At 05:00AM, the Starfish 1 team arrives at the lagoon and relieves Starfish 2. They once again give each other the thumbs up sign.

Lieutenant Cody Custer sees his Commander through his Seaglass and they catch each other's eye, "One last time, huh, sir? It sure went by quickly. This gear is fantastic. Can we keep it permanently?"

"We'll see, Tiger Fox. GREAT WORK Starfish 2!! You even got their tattoos recorded while they were swimming. NICE. I know they love that kind of stuff at biometrics."

"Sir, we moved the Sheikh's head and body out of the lagoon. It is a couple hundred yards beyond the net entrance. We just didn't want to attract sharks to this area, again, while we were working. Even with the Shark Shields, we thought it best to be careful."

"Good thinking, Tiger Fox. I know I can count on you. You are rock solid."

Lori Sanders steps out of the video wall area, just before 05:00AM to go to the ladies room. Kristi Kruger and Brian Pierce are left to observe the scene before them. They have been sleeping in shifts at the NSA, sharing the load with other analysts, but have been "on watch" the entire time.

Lieutenant Cody Custer responds to Colt Jackson, "All in a day's work, sir.

Papa Bear, I've got to be honest with you, though. Seeing all those women come and go from the beach area and courtyard has made us all a bit horny---sorry to be so crass, sir. But, I have NEVER seen so many beautiful women in my life in one place, at the same time! I must have done something wrong in my previous life, because I wasn't born into oil wealth, that's for sure!"

Colt starts laughing.

Killer Whale can't resist commenting, "Tiger Fox, I like the way you think!"

Stingray Charlie, always thinking ahead, speaks into his mike, "Yeah, Papa Bear, sir, do we get R&R afterwards?"

"Guys, when this is ALL OVER, I promise you---I'll speak directly with the boss about getting you some down time in Honolulu."

Lieutenant Cody Custer smiles through his Seaglass, "YAHOO! THANKS, SIR!"

Colt orders, "Now, get going. See you back at Ghostship in about five hours, or whenever the Muslims depart. Anyway, we should have what we need by now. I'll pick up the mosquitos at 10:00AM and we'll be back to Ghostship by noon. I'm going to swim to the jetty and place the homing-beacon hanger on the top row of boulders for the mosquitos. It looks just like the rocks piled up there, except there is a small entrance at its base, invisible from the beach."

Colt holds the rubberized "rock" up for the SEALs to see.

Tiger Fox inspects the fake rock, "Wow, it looks like the exact same color as the jetty rocks. Someone must have had some great satellite photos to make it as precise as it is. Anyway, we're on our way sir. See you back at the ranch at noon."

At 5:03AM, Kristi Kruger turns to Brian with a smile on her face. "If it's the last thing I do, I'm going to get myself to Honolulu when those guys are there! Mark my words, and YOU BETTER NOT MENTION A WORD OF THIS TO LORI!!"

Brian, hoping to join her, "Maybe we can go interrogate the people

Edward Snowden was dealing with out there. That might get us both to Hawaii!!"

"In your dreams. Those people have already been analyzed and interrogated. It will have to be something cleverer than that. I'll think about it, but I think I'm just going to put in for some personal days. We'll see what happens."

Brian looks desperately crushed by this news. He can see Kristi's infatuation and determination to meet these SEALs, these superheroes. He looks like a puppy who has just been scolded. He mumbles to himself. "*How's a guy supposed to compete with men like that?*"

Kristi looks at him, she is smiling, "What did you say?" Kristi heard every word, but is pretending not to hear anything.

Brian thinking quickly on his feet, says, "Oh, I just said "It would be great to be able to just go like that." He hopes the similar-sounding words might prevent Kristi from knowing what he truly said. He forces a smile at her, but his heart is smashed, knowing he can't compete at the SEAL's level.

Kristi just tells him, "Sometimes you just have to do, what you have to do."

At 5:04AM Bahrain time, Lori Sanders walks back into the room. The conversation between Brian Pierce and Kristi Kruger ceases and they are busy on their keyboards.

"What did I miss?"

"Nothing much, Tiger Fox was going on and on about the beautiful women there and Colt Jackson said he would be back at Ghostship by noon."

Lori Sanders looks at her watch. She calculates quickly in her head and questions her subordinates, "Why the delay? The mission is supposed to be over when he grabs the mosquitos. Those things will be in the homing-beacon rock hanger in three minutes. It only takes a little over an hour to get to the sub. Did he say he was doing something else?"

Kristi is again amazed at Lori's ability to process information quickly and analyze potential holes in any situation. Kristi gulps because she didn't even question it, she answers, "Uh...no, he didn't."

Lori looks at the video wall, "What's that man up to? Maybe it's nothing, but with Colt, every minute always counts."

On the 19th of August at 09:45AM, the Muslim leaders leave the hotel. Some of the women run up to their men and give them one last hug, goodbye. Fatima holds longingly onto General Jose Lopez for a couple of minutes. They tenderly say their goodbyes, as the Iguanas' rev up. All the leaders will be returned to the Emir's stately hydrofoil and put on private jets back to their countries.

Qatar's brand new Hamad International Airport is only five miles away and overlooks the Persian Gulf and harbor. The hydrofoil will have them all there in minutes, with a direct drop-off at the royal dock. All the leaders shake each other's hands. The Emir will be brought to the airport, as well, but by royal helicopter. Stingray Charlie captures the scene on the beach. His Seaglass records the Iguanas coming and going, as well as the landing and taking off of the royal helicopter, perfectly from his angle.

At 10:30AM Colt Jackson retrieves the mosquitos from the miniature homing-beacon hanger, disguised as a rock. All six of them made it back. Colt removes the bugs and places them on a boulder several feet above the lagoon. He checks to make sure they are transmitting their data files to the NSA. Once the transmissions are completed, they will send Colt a message, "UPLOAD COMPLETE." Then they will self-destruct.

Colt slowly wades back into the water, with only a few ripples disturbing the sun-lit lagoon at midday in Qatar, giving no indication he is even there. The Sealskin is doing its job beautifully and he invisibly slips under the lagoon. Colt rallies Chief Petty Officer Jack Flintlock and Petty Officer Duke Moran to the middle of the lagoon after putting the fake boulder in his pack, leaving no evidence behind. The mosquitos all self-terminate.

"Gentlemen, we have one last thing to do. I've brought a duffle bag. Colt unfolds it. We need to scour the floor of the lagoon for the remains of Ocean Dragon. You know what to look for---piles of crabs on decaying flesh, or bones."

Chief Petty Officer Jack Flintlock is looking seriously at Papa Bear. "I knew you wouldn't forget, sir."

Stingray Charlie looks gravely at his commanding officer, Colt Jackson. "Thank you, sir."

They all swim to the bottom of the lagoon at the point of attack and follow the trail they remember led to the vicious devouring of their brother by dozens of sharks, with one very large Tiger responsible for the kill.

Stingray Charlie spots a large number of crabs covering something below. He swims down three feet further to the bottom and pushes the crabs away. They reveal a hand, with some flesh still on it and an arm bone, nearly stripped of all flesh. The arm bone is torn off at the elbow.

"Papa Bear, Killer Whale, I found part of him."

"Ditto over here, Papa Bear. I found a foot, with nearly all its flesh eaten off of it. Ocean Dragon shouldn't have gone like this. That Emir must pay some day."

Colt sighs and tells his men, "Don't worry, if it is with my last breath, he will."

The team searches for another 30 minutes, but finds no other remains. The rest must have been eaten by the sharks.

Colt wraps up the search. "OK, men. I think we have everything we can find. Let's head back to the Seahorse."

It is now 11:00AM, Qatar time. The SEALs begin to swim back to the Seahorse. A couple hundred yards outside the entrance from the lagoon, Stingray Charlie thinks he sees something.

Petty Officer Duke Moran speaks over the COMM, "What was that?"

Papa Bear inquires, "What did you see, Stingray?"

"I thought I saw a gray figure in the water, sir. A VERY BIG gray figure!"

Colt replies, "HOW BIG??"

"It must have been over 20 feet! It has to be the Tiger, sir. He's back for the Sheikh's body they dumped in the lagoon!!"

"Killer Whale…Stingray Charlie, listen up. I'm going to drift behind you. Keep your Shark Shields on, he won't bite you if you have the signals generating. Train the hydraulic spear guns we received from Blue Angel around me. The sling-mount propulsion system and hydraulic drives increase tension in the slings. This will allow you to stay a greater distance from me."

Killer Whale asks incredulously, "GREATER DISTANCE, SIR?? What if that thing attacks you???"

"That's the plan, Killer Whale. I'm the bait. You guys get those things loaded with the cyanide darts and kill that SOB when he gets close to me. GOT IT? That's an order!! We're going to get at least some payback for Ocean Dragon."

Stingray Charlie growls, "SIR, YES SIR!"

Back at NSA in the COMM center in Bahrain, everyone is filled with tension upon hearing the plans of Colt Jackson.

Brian Pierce bellows, "HOLY SHIT, THIS GUY IS CRAZY!"

Kristi Kruger dreamingly watching the actions of the brave men, says, "No, he is just doing the right thing for a fallen brother-at-arms."

Lori Sanders thinks to herself and realizes she is a nervous wreck. All she can think is "*I'm going to kill him!*" She then says out loud. "Don't do this Colt!"

Both Kristi and Brian stare over at Lori Sanders, wondering why she is giving orders at a screen in which the other party can't even hear her.

Colt Jackson lets himself drift about twenty yards away from Stingray Charlie and Killer Whale. Both Killer Whale and Stingray Charlie aim their hydraulic spear guns on each side of Colt. As he drifts in the current, Colt turns his autonomous camouflage system OFF. He is now plainly visible in the water. He starts waving his arms and legs frantically, trying to draw the attention of the big Tiger shark. He knows it is only a matter of time until the beast arrives. Ever the one for humor under fire, and/or pressure, Colt lets loose one of the best one liners in his entire life. This breaks the tension of his SEALs and EVERYONE watching, as the potentially deadly scene unfolds.

To everyone's shock, Colt screams at the top of his lungs through his mask and headgear. Colt knows sound waves travel a great distance under water.

**"HERE, KITTY, KITTY, KITTY!...
HERE, KITTY, KITTY, KITTY!...
HERE, KITTY, KITTY, KITTY!!!"**

Chief Petty Officer Jack Flintlock starts laughing hysterically. "Sir, please STOP THAT, my ribs are hurting!"

Stingray Charlie is more cautious, "Be careful, sir. That thing's a monster."

"Get ready boys, I see an enormous grey shape, it's coming right at me! Keep your spear guns trained on both sides of me. Give him enough of a lead with your darts---and then let him have it!"

Both SEALs respond, "Aye, Papa Bear!"

In Bahrain, Brian Pierce is about ready to lose it. Kristi is biting down on her index finger and Lori Sanders is about to faint. No one says a word, except Brian Pierce.

Brian Pierce, ever the smart analyst with an elegant vocabulary, yells,

"HOLY SHIT!"

Colt Jackson can now plainly see the Tiger shark. It is definitely the same one which killed Ocean Dragon. Colt recognizes the stripes on its side. He thinks, "*Game on!*"

The only problem is, everyone else can see this dramatic confrontation, also. Admiral Steven Whistler is completely pissed off.

On August 19th at 11:14AM Qatar/Bahrain time and 04:14AM Washington D.C. time, Admiral Steven Whistler screams over the SEAL COMM line, "Papa Bear, what in the HELL are you doing! I ORDER YOU TO STAND DOWN AND TURN ON YOUR SHARK SHIELD IMMEDIATELY!! DO IT NOW, SAILOR, OR YOUR FIRED!!"

Colt Jackson, with a slight delay calmly responds to his friend, but one of the top brass in the Pentagon, "Can't talk now, sir. But, I hear you. 'All for one and one for all.' We never leave a fallen comrade behind. ***You taught me that, Polar Bear.***"

Admiral Steven Whistler has now walked into his War Room and forgets he's on the open COMM line. Everyone hears him scream.

"If he's not eaten alive, I think *I'll* kill him!"

At the Bahrain COMM Center, Lori Sanders whispers to herself, "*You'll have to beat me to it, Admiral.*"

The shark is swimming at flank speed---now only about thirty yards away from its target---Commander Colt Jackson! The 2,000 pound monster carries the same physics, power and force as a speeding motor vehicle. In this case, it won't only run right over Colt Jackson, but will devour him!

The mammoth Tiger's jaws begin to open, revealing the set of 4-inch, razor-sharp teeth which killed Ocean Dragon. Colt's eye movement to the shark's mouth causes the Seaglass video camera to automatically zoom in on the approaching shark's gaping teeth. A vivid picture of multiple rows of enormous teeth shows up on everyone's video wall. Audible gasps are heard in every COMM center monitoring the mission.

The people in the COMM rooms at the NSA in Bahrain, at Ghostship, at Ft. Meade and at the Pentagon stand petrified with fear for Commander Colt Jackson.

The shark is approaching Colt in a direct, frontal attack! It is not approaching from either side of Colt's body. If Killer Whale, or Stingray Charlie, takes a shot at the Tiger with their spear guns, they will hit Colt.

The only way for them to reach the enormous predator and kill the shark, is to shoot right through Colt's back. Colt realizes his predicament and, thinking quickly, he pulls out the only weapon he has---his trusty old, Microtech DOC Flipper Knife!

Colt and the Tiger shark look at each other's eyeballs. The two-thousand pounds of man-eating fish, speeds closer to Colt. The Tiger widely opens its jaws, ready for the kill. Its eyelids begin to close when it is just yards away from Commander Colt Jackson.

Lori Sanders, not caring who hears her says, "Oh, Lord, please protect

Colt!! PLEASE. Don't let him die. PLEASE!!"

Brian Pierce kicks it up a notch and screams even louder, "HOLY SHIT!!" Kristi Kruger fearing for Colt, but admiring his raw courage, says, "What a man! I'm glad I had a chance to meet him. I don't think I can watch this." She shuts her eyes and bows her head. She may be praying, but it is hard to tell. Lori doesn't know Kristi's religious persuasion, if any.

Colt, knife in hand when the Tiger shark hits the five yard mark, simultaneously flips a switch to turn the Shark Shield ON! The enormous shark immediately tries to veer off, as the incredibly strong electronic impulses from the Shark Shield warn the shark of danger.

The impulse waves are lighting up the Tiger's nostrils with powerful electric shocks absorbed directly by the sharks Ampullae. This creates the feeling of lightning in the shark's snout and body. The shark swerves violently to Colt's left, trying to flee the threat.

Stingray Charlie, holds his spear gun ready, he knows he is covering this side of Papa Bear's body. Colt knows about sea life and he also knows a shark's most feared enemies. Other than man and a killer whale, it is the dolphin. Dolphins use their bottle noses to kill sharks by hammering them in the gills.

As the shark continues its violent turn to Colt's left, its gills are a foot from Colt's face. Colt pulls out his DOC Flipper knife.

In Bahrain, everyone in the room is gripped with fear at the scene. Lori Sanders eyes start to water. She can't help it. Kristi feels the stress in the room and she opens her eyes just in time to see Colt plunge the DOC Flipper Knife into the shark's gills.

Brian Pierce is about to lose it, "HOLY SHIT!"

Kristi Kruger cries out, "UNBELIEVABLE. WHO IS THIS GUY???"

Lori Sanders is praying, "OH, GOD, PLEASE KEEP HIM SAFE!!"

In the Pentagon, Admiral Steven Whistler says, "Kill him, Colt." The Admiral forgets to use Commander Colt Jackson's call sign "Papa Bear."

After watching this unbelievable scene play out, Brian states flatly for everyone to hear, "That man has the biggest pair of balls I've ever seen in my life."

Lori looks sternly at Brian, raising her eyebrow, "Brian Pierce...SHUT IT!! DO YOUR JOB!! Do we have all the uploads yet?"

Brian points wildly at the video screen. "But, boss, WATCH!!!"

Colt is using the same knife he used to save Prince Harry in Afghanistan. It is also the same exact knife he used to kill the criminal merc from Russia at the Federal Reserve in New York, last year. Colt pulls the knife out of the side of the beast's gills.

Colt Jackson is filled with adrenaline and screams, "Take that you miserable bastard! And take this, again, for Ocean Dragon!!" Papa Bear plunges the blade once more into the side of the Tiger.

Such force was used with the last plunge of the knife the video camera on the Seaglass captures a clear shot of only the top two inches of the handle on the knife, protruding from the shark's huge body. The rest of the weapon is buried deep in the side of the Tiger. As this eating machine swims away, a trail of blood pours out of the beast.

Killer Whale says to Petty Officer Duke Moran, "You've got the ball, Stingray Charlie. The Tiger's swimming out on your side. Finish the job."

"Roger that. I'm locked on target. Fire away."

The poison dart from the hydraulic spear gun is aimed five yards to the left of Colt Jackson's back. As the shark completes it turn away from the Shark Shield and Papa Bear, the dart finds its target. It is a perfect hit, right behind the gills and in front of the knife of the great Tiger shark.

The poison dart is propelled with the hydrodynamic force found in the new spear guns. It finds its target perfectly and burrows deep into the shark. The cyanide immediately takes effect. The huge twenty-foot Tiger shark plunges to the bottom of the lagoon, below.

Inside the USS Minnesota, a crowd is gathered in the CSRR. The entire communications room is packed with sailors.

They all witness the scene of Colt Jackson luring the shark in and then taking it out with his Microtech DOC Flipper Knife. When the shark is hit by the two knife thrusts, the sub erupts in huge cheers. When Stingray Charlie connects, on an over twenty-yard long spear gun shot, both Stingray Charlie and Papa Bear become instant legends on Ghostship.

On their video wall in the sub, they can see Colt Jackson as he swims down to the big fish, still wreathing on the bottom of the lagoon. It is turned over on its back, with its stomach turned up. Colt straddles the gigantic Tiger like a horse. He looks over the shark, and feels for his lucky DOC Flipper knife. The shark is still wriggling beneath him.

Colt shouts at the Tiger shark, "Not so tough now, are you pal? NOBODY, AND I MEAN NOBODY TAKES MY DOC FLIPPER KNIFE! It's saved my life so many times, I think I've lost count!" Colt forcefully grabs the handle of the knife. It is barely sticking up out of the side of the shark. He yanks it out of the shark's side.

Colt then violently plunges his lucky knife into the belly of the shark. Brute strength and adrenaline allow him to begin slitting the belly wide open. It is a gruesome site.

When Colt opens the belly of the beast up wide, inside are the remains of the corpse of Petty Officer Paul Westwood. The scene captured digitally and recorded by the Seaglass, is horrific. Everyone at the NSA BAHRAIN center gasps. It is an ugly and scary site they see. The partially digested remains of the head and torso of a very brave U.S. Navy SEAL---Petty Officer Paul Westwood, AKA Ocean Dragon---is too grotesque for the analysts to watch.

Everyone gasps in the NSA center. Some put their hands over their faces, some put their hands over their mouths, and some turn away from the horrific video being captured by Commander Colt Jackson's Seaglass.

Collectively, they all catch their breath is shock,

"HUUHHH!"

Kristi Kruger is hiding her eyes. She says, "I think I'm going to be sick."

Brian Pierce is wringing his hands together and says, "What a horrible way to go."

Lori Sanders' hands are on her cheeks. She looks away. "God Bless that brave SEAL."

General Shane Maxwell, leader of all of the NSA, tells Commander Jackson over the live Seal COMM, "Papa Bear, this is Four Eyes. I will never forget what you did for your fallen brother, Ocean Dragon. Great work, son."

The General continues on a private, encrypted line to Admiral Steven Whistler. Papa Bear cannot hear this private communication. Now the General is chuckling a bit while he speaks to the Admiral. "Polar Bear, I'm sure you're mad as hell at Papa Bear, but all I can say is, who do you think he learned that from? Sorry to laugh when the remains of a brave warrior are on the screen, but those who rescued his remains remind me of another crazy-assed U.S. Navy SEAL about fifteen years ago. You do remember him, don't you Polar Bear? You did nearly the EXACT SAME THING when you rescued my ass from that tiger in Southeast Asia. You DO remember, don't you POLAR BEAR??"

"I remember, and thanks for the reminder, Four Eyes. I'll go easy on him, no worries." The Admiral continues on with Papa Bear, still somewhat angry. "Good work, Papa Bear. We are all proud of your team and how you retrieved Ocean Dragon. WE DO NOT LEAVE MEN BEHIND! Now, get your butt back to Ghostship ASAP and report in. We need to talk, OFF THE COMM!!"

Colt Jackson is still on the floor of the Persian Gulf and says, "Aye, aye, sir. But, I obeyed your order, please remember that, sir. I just waited a few seconds to execute it..."

The Admiral is shaking his head and says, "We will need to discuss tactics in a little more detail for the next mission, soldier."

"Polar Bear, with all due respect, sir. I did follow the rules of engagement 100%."

"How's that, Papa Bear?"

"Well, you specifically ordered us not to engage, unless we were engaged upon, sir. I'd say a massive twenty-foot, man-eating Tiger shark attacking me on a beeline, ten yards away, coming at flank speed, qualifies as an engagement! I was merely following orders, sir."

"Don't be a smart ass, Papa Bear. Thanks for bringing our boy home. Speak to you when you return to Ghostship. Over and out."

Colt Jackson, Chief Petty Officer Jack Flintlock and Petty Officer Duke Moran retrieve the remains of Ocean Dragon and swim out of the lagoon. They reverently and gently escort the body of Petty Officer Paul Westwood, cradling it between them.

At 11:45AM, Lori Sanders has now got the ball. As head of Internet PRISM operations for the NSA, she is directly responsible for interpreting the audio and video feeds from the mosquitos, received over the encrypted Internet channel. She will have to prepare a briefing for Admiral Steven Whistler, General Shane Maxwell and, depending on the gravity of the information, the President of the United States. Lori can be as tough as nails when she needs to be. Now is such a time.

"Do we have the entire upload sliced and diced yet? Get on it people! This is critical to national security. When you have it, queue up mosquito 1's audio and video footage from August 17th from the Grand Ballroom beginning at noon. This is when the big meeting started. Got it?"

Kristi Kruger replies, "Yes, ma'am."

Brian Pierce says, "We're on it."

Lori continues, "I've split the rest of you into five groups. I want every second of the other five mosquitos' footage analyzed and tagged. Use the audio and video speed-up capability to accelerate your analysis, when there is no action. TIME IS OF THE ESSENCE." She claps her hands quickly and in succession. "NOW MOVE IT PEOPLE!!"

There is a mass of activity with the NSA teams gathering to begin the analysis at their respective stations.

At 12:45PM, Colt Jackson, Chief Petty Officer Jack Flintlock and Petty Officer Duke Moran return to the sub and dock the Seahorse. When they leave the lock-out chamber and enter the interior of the sub, they are mobbed by the crew of the USS Minnesota nuclear attack submarine.

The "selfies" go on for over three hours. Each and every crew member wants his picture taken with all three heroes. Cell phones flash continuously.

CHAPTER 9

THE DREAMLINER

By 1:00PM, after the meeting had broken up at the Marsa Malaz Hotel at the Pearl, General Jose Lopez and Emir Sheikh Muhammad Al Therani, ruler of Qatar, board the Emir's private Boeing 787 Dreamliner.

Being constantly in the presence of the other Islamic leaders, and now the Emir, has prevented the General from contacting anyone at the CIA, or NSA. He knows the events which occurred at the hotel during the last several days have critical security implications for the entire world. He thinks, "*I must get a report back to the Admiral.*" He only hopes the mosquitos did their job. He decides when they land in Caracas, the first chance he gets, he will phone in and report directly via his encrypted phone.

The General looks around the plane. The interior of the enormous aircraft has been customized with lambskin leather seats. They are soft and smooth to the touch. All the trimmings are in solid, 24-carat gold---except the toilets, where the wash basins, faucets and fixtures are "only" 18-carats. 24-carat gold was thought to be too soft for the wear, tear and constant usage, a bathroom experiences.

The distance between D'oha and Caracas is 7,624 miles. The Dreamliner speeds along at 593 miles per hour. Their flight time, against a 100-knot headwind will take them approximately sixteen hours. Given the time zone changes, the General calculates they should arrive at Simon Bolivar International Airport at 9:30PM local time on this day, August 19th.

In the meantime, the Emir's Dreamliner has everything and anything General Jose Lopez can imagine. He looks around and sees belly dancers, a pool, an exercise facility, a cinema, a state room, a private dining room, a dozen different chefs and literally the choice of any food and drink imaginable. The Emir has also brought along his harem. Before retiring to his private chambers, the Emir puts his arms around six virgins. Each of them is exquisite and he has saved these women for something important.

The Emir is just about to depart for his private quarters, but engages the General in one last conversation before he retreats with his women. By the look on his face, the General can see it is something important.

"General, I am going to retire to my private chambers and get some rest, if I can. Sometimes it is difficult to rest when Allah blesses us with such beautiful creatures, like these lovely women around me."

The members of his harem giggle. The Emir continues, "But, I have a special gift for you. I leave you with these six, amazingly gorgeous virgins. They will wash you, they will play with you, and they will be yours. They will comfort you in any manner you desire."

The General is stunned. "Thank you, my Emir. And *thank you* for planning such a fantastically successful summit for our leaders. This is the plan, all Muslims will remember, which broke the back of the Great Satan---AMERICA!"

Emir Sheikh Muhammad Al Therani reaches toward heaven, "Insha'Allah. Allahu Akbar!" (May God will it. Allah is the greatest!).

The General responds in kind, "Insha'Allah! Allahu Akbar!"

The Sheikh then explains to the General, "I didn't tell the others you and I were departing to Venezuela, together, and I appreciate you keeping this confidential---at least on a temporary basis. I did not want our leaders distracted on the eve of our attack on the West!"

"I completely understand, my Emir!"

"After the attack on America, South America will be ours for the taking! Imagine hundreds of millions of Catholics forced by the blade to convert to the will of Allah!"

The General replies, "Allahu Akbar, great one!"

"It is for this reason I wanted to meet privately with you and your forward-thinking President in Venezuela. We need to grow close relationships with him---before he is overthrown and you are put in his place. I have selected you to be the Emir of Venezuela and the leader of ISSA, when our victory is complete. "

"ISSA" is the Islamic State of South America.

General Jose Lopez smiles broadly and bows his head to the Sheikh, "My Emir! I don't know what to say. I have been blessed today by Allah. Thank you for your faith and trust in me! Allahu Akbar!!"

"Allahu Akbar, General. And don't worry about affording the six virgins. I have already wired into your numbered account in Singapore $5 Million. Until you are named Emir, this will afford you the ladies. I know your government doesn't pay its leaders much!"

The General truly doesn't know what to say. He looks around at the beauty of the aircraft and then looks into the Emir's eyes, "I am truly blessed this very fine day, indeed, my Emir. I will not let you down and I will continue to do the Prophet's work!"

"Your launching the EMP attack from the oil tankers over Texas, California and South Carolina, will make you a Prophet Warrior. Potentially, you could become much more powerful than even our great Caliph Abu-Bakr. You have a special gift of being able to blend in with people. Today, on this flight, you will change your stars. Welcome to my world, my future Emir!"

The General can see his star is rising. He smiles graciously and is stunned by these plans of the Emir. "Your will is my will, great one. To Allah be all glory. Allahu Akbar! Pleasant dreams, my Emir, my great leader!"

The Emir smiles and says, "I hope you don't sleep a wink! Have a good "rest." The Sheikh smiles brightly at the General. He turns and is escorted to his chambers by a dozen other beautiful women. The great Sheikh leaves the General to pursue his own desires.

General Jose Lopez is in shock. He did not expect this from the Emir. The Emir is setting him up to be a Caliph! He now looks at the six, beautiful virgins he has been given, as gifts by Emir Sheikh Muhammad Al Therani. They are all veiled, but he can see they have voluptuous figures. He looks them up and down, from left to right. When he gets to the last woman, they make eye contact. The General stares intently. The virgin slowly and invitingly, lowers her veil. It is Fatima! The General's heart, races.

By 09:50PM, the Emir's Dreamliner lands, twenty minutes later than scheduled. "Simón Bolívar" International Airport is located in Maiquetía, about 13 miles from downtown Caracas, the capital of Venezuela. The airport is located at the northern base of the luscious, steep-green mountains of Venezuela's coastal range. It is called the Cordilla de la Costa and rises above the Caribbean Sea.

Caracas, itself, is overlooked by Mount Avila and is built in a steep and narrow valley to the south. The city is surrounded by mountains and hills. Before the invention of the automobile, the city was especially beautiful, but, today, smog can be a real problem.

The Cordilla de la Costa is the city's best friend and mortal enemy---the pollution often gets trapped in the valley by the mountains surrounding it. The air pollution index, according to numbeo.com is 62.5 compared to New York City, which is 51.79 and Los Angeles which is 56.25.

As the General, the Emir and his entourage deplane, they are greeted with a full military honor guard. Their black, bullet proof SUVs await them.

They will be escorted to the Renaissance Caracas where they have taken the top floor, which includes the Presidential suite. It is barbaric by the Emir's standards, but would be considered exquisite by the rest-of-the-world.

General Jose Lopez walks with the Emir and they shake hands, giving each other the customary Middle Eastern hug. The General will be escorted to his home by military convoy. Guerillas are everywhere in Venezuela and no one can be too careful. Crime, extreme poverty and the difference between the relatively few rich and the millions upon millions of poor people, make this country a showcase of plunder by the Spanish empire.

Emir Sheikh Muhammad Al Therani says goodbye, "Until tomorrow, then, my General." He extends his hand to the General.

"Yes, my Emir. I will meet you at the Presidential Palace at twelve noon. President Maduro is looking forward to feasting with you."

The Sheikh responds, "Until then, my friend."

They depart in their separate convoys.

In the military convoy, the General is riding along with his staff.

"Take me to my home between Los Palos Grandes and Alta Mira. I will sleep there, tonight, instead of my home at Las Mercedes."

The General's driver and attaché quickly complies, "Yes sir, General."

General Jose Lopez lives in luxury. He has a beautiful, gated and guarded mansion in Palos Grandes. It is still the rainy season and it is hot and humid outside. The General goes to his safe, dials the combination, and pulls out his encrypted phone. He opens the sliding glass, bullet-proof door and walks outside on the back terrace, just to make sure his conversation isn't being recorded. He leaves his six virgins inside, with Fatima looking at him, longingly.

The General slaps at real mosquitos buzzing around his head. He picks one up after killing it. He inspects it, carefully. It looks nearly identical to the tiny miniature robots Colt Jackson gave him. "*Amazing*," he thinks.

Then he speaks to himself, out loud, "Increible!" It looks just like the nano-tech mosquito Commander Jackson gave me. What will those Yankees think of next?" He looks at the keypad of his secure, encrypted phone and dials the Admiral. "Hello, Polar Bear, this is Croc Hunter, over."

Admiral Steven Whistler, relieved to hear from the General says, "Croc Hunter---we were getting worried. Is everything all right?"

The General explains, "I was surrounded the entire time. The Emir is meeting with President Maduro and me, tomorrow. We flew to Caracas on his personal Dreamliner, immediately after the summit broke up. There was no way I could make a secure call at any minute, until now. I just arrived back home. How much do you know? Did my six little friends report in and do their job?"

"Affirmative, we're still analyzing all the information, but we have the summit nailed. I don't need to tell you what is at stake here."

The General briefs the Admiral on his perspective, "They are dead serious and have the tools to accomplish this dreadful mission. This I can vouch for. It is, how can I say it? Muy peligroso (Very dangerous)."

"Te acuerdo, senor (I agree with you, sir). Did you learn anything from the Emir? I don't know how to say this, sir, but it is extremely difficult for me right now. The Emir has plans to make ME the Emir of Venezuela, after the EMP attack is made on the U.S."

"He does, huh?"

"*Polar Bear*, I am to be the Emir of Venezuela and the Chief Cleric for ISSA---the Islamic State of South America!"

The Admiral whistles into the phone, "WHAT? Anything else??"

"Well, he has already wired $5 Million to me. Sir, there is something else. I'm not very comfortable talking about it. You know I am a moderate Muslim and I have, what I thought, were high moral standards."

"That's why I picked you for this journey you're on. You are intelligent, special and sensitive to the moral law which rules, truly, every man's heart, Croc Hunter."

"Sir? Can I speak off the record?"

"Of course."

"I just feel corrupted now. To keep my cover at The Pearl, I slept with three different women and participated in an orgy. Then, when I boarded the Emir's plane, a half-hour into the flight, he presented me with six virgins---which he insists I now own. They are all inside my home, as we speak! He said he knew I didn't make enough money to support them in the style he thought appropriate, so he put the $5 Million into a numbered Singapore account."

Admiral Steven Whistler absorbs this information. A pregnant silence follows. "Are you OK, Croc Hunter? I have to tell you, tradecraft can be very cruel and can shatter your heart. We all have had to do things for the greater good, which are disdainful to us. The question is, are you being bought off? Does the Emir know something about our relationship?"

The General replies, "I can guarantee you, if he even had a whiff of our relationship, my eyeballs would be looking up at my torso because the sharp blade of a scimitar would cut my head off."

"Yes, we were all aghast at the pure brutality of the murder of the Sheikh. Croc Hunter, *can we count on you?*"

"Without question, sir. The best way forward is to allow the three tankers I have waiting in Maracaibo, receive the North Korean cargo offload from General Kim Se Bong. It will take the North Korean cargo ship a total of thirty-seven days to reach Maracaibo. Right now the ship is going through the Suez Canal. The General thought this route was less risky than going through the Panama Canal. They are flying a Panamanian-registered flag, to also reduce the risk of getting caught. He has already paid everyone off in Egypt to get through the canal."

The Admiral is typing on his keyboard. "Let's see…that's 13,619 nautical miles. With an average speed of 15 knots, it will take it 37 days and 20 hours to arrive at port.

"In one sense, it is perfect for follow through from the summit. This gives us enough time to prepare the swimming pool encasings and the cement/lead barriers for the domes we will place over the missiles and their nuclear payloads. The tankers I have reserved in Maracaibo are from three different flag states---Panama, the Marshall Islands and Singapore."

"This sounds good, Croc Hunter. You have to keep up the appearance and expectations built up from the summit. You never know who will show up in Maracaibo to inspect the ships. The Emir's and the Islamic State's resources are nearly infinite. They will undoubtedly have spies on the ground, overseeing and reporting what is transpiring during the next month-and-a-half. Let me look at the registry data for a second."

The Admiral again types into his keyboard. "Yes, those flagships are the three most popular flags in the oil-tanker market. Out of a total of over 4,295 tankers, Panama has 528, the Marshall Islands have 234 and Singapore has 355. They should blend right in. We have to make some serious decisions about whether, or not, to intercede when they finally get set to launch the missiles."

"What do you mean, Polar Bear? You MUST intercede!"

"I'll come back to you with the plan, Croc Hunter."

The General confirms the dates, "Everything is set to "go" for Yom

Kippur, the Jews' holiest day of the year. It's on October 11th this year.

"Yes, we picked that piece of Intel up from the summit. Croc Hunter---one last thought, sometimes God uses us to do certain things which in themselves, might be considered shameful. The shame is not on you. Please remember this. The shame is on those we must penetrate by pretending to be like them---in order to accomplish the mission and potentially save millions and millions of peoples' lives. Don't ever forget this."

"Thank you, Polar Bear. I needed to be reminded of this. The temptation is something else, sir."

"I can only imagine. Six virgins? Stay focused on your job and remember, assume they are still under the Emir's control. They'll probably report anything strange---like you not sleeping with them. You might not be the next Emir of Venezuela, but I think we can make you its next President."

"Thank you, sir. Thank you very much. But, I consider what I am doing to be my duty as an American and a patriot---that is enough for me. I also strongly disagree with the radical Islamists destroying the religion of Islam for nearly two billion people on earth."

"You're a brave man, Croc Hunter. Get some rest. I'm sure you're exhausted. We'll touch base in September, or sooner, if something else occurs. Keep your ears and eyes open."

"Aye, aye, sir. One last question, sir. What do I do with the $5 Million?"

The Admiral doesn't hesitate, "Spend it to advance your cover and set yourself up to replace President Maduro. Keep your friends close and your enemies closer, Croc Hunter. This one is for all the marbles."

"Aye, aye, sir. Over and out."

CHAPTER 10

THE RAPTOR

At Peterson Air Force Base in Colorado Springs, Colorado on Tuesday, August 23rd at 2:00PM Mountain Time, Air Force One lands. President John Diaz steps off the stairway of the plane and is greeted by the Commander of NORAD and USNORTHCOM, General Charles Jacobs.

General Jacobs is a United States Military Academy graduate. He has distinguished himself in the U.S. Army, primarily, throughout his career. He commanded the First Corps, including a combat tour in Iraq serving as the Commanding General of the Multi-National Corps-Iraq.

He got to know the President when he worked as Director of Strategic Plans and Policy, "the J5" for the Joint Chiefs of Staff. He also has the full confidence of Admiral Steven Whistler as a result of his work with the Joint Chiefs. General Jacobs is also no pansy. He was ex-Airborne and spent a great deal of time at Ft. Bragg, North Carolina. The President likes him. When the President looked into his background, he discovered General Jacobs held a Master's degree in History from the University of Michigan---also the President's Alma Mater.

When Commander Colt Jackson recommended the plan to replace U.S. Federal Reserve Notes with a 21st Century version of Abraham Lincoln's Greenbacks, it was General Jacobs the President called as a sounding board. The General had wholeheartedly embraced and encouraged the President to implement Commander Jackson's plan. By implementing this plan and attacking the Fed, the President had ended the financial crisis of D'Apocalypse™ and the pending economic collapse last year. It also made him one of the most popular Presidents in history.

By solidly supporting Colt Jackson's idea and giving inspiration and confidence throughout the military ranks, supporting the President at a very delicate time in the nation's history, the General made a friend of the President, for life. The President believed his strategic planning capabilities would make him an excellent fit to run NORAD and USNORTHCOM.

When the President informed the General he was going to personally recommend him for the job, General Jacobs had protested, saying he wasn't qualified because he was not an Air Force officer. The President countermanded the protest and told him to surround himself with the Air Forces' very best. Since this time, the General has been "on watch" for nearly three years. During his tenure, he has advanced revolutionary defense strategies, tactics and technologies, which have been deployed to protect America.

"Welcome, Mr. President. It's good to see you, sir."

"Hello, General. How are you enjoying this 75-degree September day, with no humidity and all this sunshine in Colorado Springs? Have you been able to golf much at Broadmoor? I hear it's beautiful. In all my years, I have never been there. I'm looking forward to meeting the press later in the day and if I can, playing a quick nine holes."

"No sir, I'm not a golfer. But, we do have the top floor of the Broadmoor cordoned off for you, in the event you choose to stay overnight."

"I'd love to play. I guess we'll see how much time the meeting takes. How long will it take to get there?"

"Only minutes, sir. In case you have the time and need a couple of hours to clear your head, the course comes right up to the hotel. In any event sir, we're going to be leaving this facility. Have you been on the Maglev, lately?"

The President looks at the General and says, "Only to get to Mount Weather for some emergency planning. I find it rather amazing how talented our intelligence community is."

"How's that, sir?"

"They are just downright brilliant, in a more important way than just being smart in developing technology and deploying it."

"And how's that, sir, if you don't mind me asking? "

"It's simple. Their greatest strength is disinformation! They focus attention AWAY from secrets. For example, when China makes big announcements about advancements in their nuclear programs, like their highly touted thoreum-based and pebble-bed based gas reactor technologies, we keep SILENT. This leads the world and EVERY reporter in the world, foolishly thinking we have dropped the ball! This, despite the fact, we invented these technologies in the 1960s at Oak Ridge, Tennessee! I mean, General, you run NORAD now. You have those miniature thorium-based reactors all over the place in the underground city. When NORAD digs super-deep with the nuclear drilling machines, you use hydrogen-based gas reactors because they don't require water! They aren't bothered by the temperature increases as you drill deeper into the Earth's crust."

"Of course sir, I see your point. All of our training prevents us from reacting to any news from our enemies."

"Well, all of you do a great job of it, General."

Air Force One landed only a short walk from the building on Patterson Air Force base, which serves as the public headquarters for NORAD and NORTHCOM. The whole world knows, however, the real action takes place inside Cheyenne Mountain---and the miles of tunnels hidden beneath, and around, it.

The President walks by a plaque prominently displayed on the headquarters' building. It reads as follows:

EBERHART-FINDLEY BUILDING

Named in Honor of General Ralph E. Eberhart, USAF, Eighteenth Commander, North American Aerospace Defense Command and First Commander, United States Northern Command, 2000 - 2004, and Lieutenant-General Eric A. Findley, CF, Battle Staff Director, Director of Operations, and Nineteenth Deputy Commander, North American Aerospace Defense Command, 2001 – 2007.

THEY HAD THE WATCH

The General sees the President reading the plaque. He further explains to the President, "General Eberhart and General Findley had the watch on 9/11---Sept. 11, 2001. As you know, sir, the country was in chaos. These men created a military command devoted to solely guarding the homeland. Together, they revolutionized our thinking and created a new geographically-based command which united Canada and the U.S. As I'm pretty sure you know, sir, General Findley is Canadian."

"Yes, I am aware of this. It is good to see them honored in this way."

"Sir, you haven't visited here since you pressed for my appointment to run NORAD, but I have something interesting to show you. I didn't brief you on it yet because the project was only running on a very small-scale, on an unproven, test basis."

The General and the President have been talking as they walk into NORAD's lobby floor entrance. The General presses the down button on the elevator. The General and the President drop quickly---26 levels underground. The speeding elevator gets there in no time. When they come out of the elevator, there is an active city of moving people and brightly lit hallways. A big sign greets them as they leave the "-26th" floor. It says,

2,203 FEET UNDERGROUND---WE HAVE THE WATCH!

"I believe your Maglev from the White House to Mount Weather runs a little over 300 miles per hour on the magnetically-driven levitated train. You cover the 60 miles in what?...10 minutes max, between stopping, starting and boarding/de-boarding?" the General asks.

U.S Air Force Tunnel Boring Machine

Work product of U.S. Government
Public Domain Image

"Yes, that's about right. I don't even have time to have a cup of coffee, it's so fast."

"Well, sir. We have just deployed the first legs of the Super Maglev technology. By using the nuclear powered tunnel-boring machines, which melt solid rock and leaves behind glass-like walls, we were able to carve out perfect tunnels to further reduce atmospheric pressure. This dramatically increased the speeds of the trains."

"How fast are they?"

The General sheepishly looks at the President and says, "Well, sir, we've had them going at nearly 1,800 miles per hour on the long tunnel between here and Area 51. It takes about 20 minutes. It's a 600 mile trip."

"Are you serious?"

"I am sir. We have also created a name for them. We christened the trains, 'VRAPTORS.'"

The President laughs. "You guys crack me up with your nicknames and call signs. You are always creative. OK, I'll bite, what does "VRAPTORS" stand for?"

"It stands for "Vacuum Reduced Air Pressure Train Operating Rail System." VRAPTORS. Cool, huh?"

"Very cool General, but why not just call them "RAPTORS?" It's sort of like "The Facebook" being shortened to "Facebook." It's catchy. When is it coming to the White House?"

"That is WAY cooler than VRAPTORS, sir. If you are OK with it, we will rechristen them 'RAPTORS' in your honor."

"By all means, General!"

"Sir, we're operational, already. We did suffer through one incident, however. While one of the RAPTORS was operating, a tube section failed and leaked. This caused the air pressure in the system to vastly increase---which caused a massive and sudden de-acceleration of the train. It put about a dozen people in the hospital. There were no fatalities though, because everyone was wearing a safety pod and were all seated."

The President looks at the Admiral, "Good thing no one was walking around!"

"We have to think through the safety implications, but this was actually a great test of the system. The leak was caused by a defect during manufacturing. We checked the entire track line with a fine-toothed comb and there are no more leaks. The RAPTORS are now working like a charm. During the last two months, alone, this has accelerated our technology exchange with U.S. Space Command and their special visitors at Area 51."

The President is thinking, "What happens if an earthquake hits? Won't this cause a disaster and huge costs to repair the system?"

"Sir, the system is built with an ultra-flexible, ultra-tough material. It is the hardest, yet one of the most flexible materials, in the world."

"What is it?"

"It's graphene. It consists of a single atomic layer of carbon, with a natural honeycomb structure."

"I used to own a pair of Hexcel aluminum, honeycomb-cored skis. They were incredibly strong and flexible."

"Using graphene and a similar design makes the tunnels virtually earthquake and nuke-proof, sir.

President John Diaz stares at the General. "That's remarkable."

The General continues, "The thermal stability of graphene is extremely high, its melting temperature is possibly the highest in nature. It melts at 4,900 Kelvin…It's weird stuff, that's for sure. We even have a supercapacitor able to store electricity, using graphene, at -58 degrees Fahrenheit. In our Ann Arbor days, I'll bet we both could have used a battery built out of graphene during the winters back there."

"I still remember how numb my fingers were, freezing in January---brrrrr!"

"We made a sizeable bet on graphene, sir, and it's paying off big time. We worked closely with Rice University and Georgia Tech to work out the initial problems with large sheet imperfections in the material. This, it was determined, was the reason the tube failed."

"I knew you were the right man for the job, General. We don't just live in a brute force world, anymore. Intellectual elegance is as much a part of military and intelligence strategy as raw strength was to winning the Cold War against Communism. I have a question. Right now we're a little over seven football fields below the surface. We were at 75 degrees on top. At 2,203 feet below the surface, how hot is the external rock temperature down here, anyway?"

The General points over to the wall. Two thermometers show the temperature, inside the city, as well as on the other side of the city's wall, inside the earth's crust. "It's 114 degrees Fahrenheit outside in the rocks around the city, give or take!"

"It doesn't feel the slightest bit hot down here. It feels like the exact same temperature as the outside air at the surface. A perfect 75 degrees."

"Thanks to the thorium reactors, we have an unlimited supply of electricity. You are right about the agencies. They are very clever bastards, over there, sir. Forgive the language."

President John Diaz is laughing. "I've heard far worse, General. Don't worry about it. I was briefed on Thorium. It is plentiful and it will not reach critical mass, where a runaway meltdown occurs. If it begins to overheat, a plug melts and the salts drain into a pan and the salt mixes with the fuel. This spreads it out and allows the system to cool down on its own. The system is supposed to be bulletproof against a radiation event and there is about 1/1000 the amount of nuclear waste compared to a traditional uranium reactor."

General Jacobs looks with surprise at the President's firm grasp of knowledge on the workings of a Thorium reactor. "That's correct sir. In an emergency, or God forbid, no human beings were down here, or somehow, the tunnel system was completely destroyed, the reactors will shut down automatically without even needing a computer---or electricity!"

"We're going to push these reactors to the forefront, General. It's about time. We need to get off fossil fuels."

"With other countries now actually using the technology, sir, there is no reason not to unleash our innovations on the civilian market. Here we go, sir." The General has walked with the President to the entrance of the RAPTOR station. The General points the way. "We'll be at Denver International Airport in just a couple of minutes, sir. We had you fly to

Colorado Springs and everyone else to Denver. This helps avoid the Press from connecting the dots."

"Good work, General. Let's go! The RAPTOR awaits!" They depart on the RAPTOR rail system.

CHAPTER 11

ANGEL SHIELD

Denver International Airport covers an area of 54 square miles. It is the largest airport, by area, in the United States. The Denver Jeppesen Terminal is immediately recognizable by its white peaked, fabric roof, which the architects stated were designed to remind people of the snow-capped Rocky Mountains directly to the west, as well as the Native American teepees which used to populate the western landscape.

The airport is the source of countless conspiracy theories. In the baggage area are strange, Armageddon-like murals. Near the murals are the letters, "AU AG," which are inscribed in the floor. Many believe this stands for Australia Antigen, which is another name for the acute illness Hepatitis B, which is rarely fatal, provided a vaccination is taken. Hepatitis B causes liver inflammation, vomiting and jaundice. The disease has a poor response to most treatments and there is speculation these images of death in the murals reflect various ways in which people will die in the end times.

The airport received thousands of complaints about their morbidity. As a result, many of the murals were removed, or painted over, because of the extreme concern over the satanic symbolism present in the artwork. One of them even featured a Nazi-monster figure, holding an AK-47 in one hand and a scimitar in the other, killing the dove of peace. Another showed three different females, all dead in caskets. These included a Judeo-Christian, an African-American and Native American woman. In this mural, cities and trees are burning in the background, while a strange penguin is protected in an ice-box. The Aztec's Quetzal bird was married by the Aztecs with a snake. It became their "feathered serpent"---Quetzalcoatl.

Quetzalcoatl was related to the Aztec gods of the wind, the planet Venus, sunrise, as well as business, arts, crafts and knowledge. He was also the patron god of the Aztec priesthood. Snake venom was used by the priests as a prescription for many treatments when people fell ill.

The existence of the belief in Quetzalcoatl and other gods is one of the key reasons the Spanish destroyed the Aztecs, Mayas, and Incas in Central and South America.

Quetzalcoatl was one of several important gods to these cultures. All of them had a clear history of human sacrifice, unspeakable torture and murder which took place at the hands of the priests. Many Christians believe Quetzalcoatl is a satanic symbol.

Aside from the murals, grotesques (fantastically ugly statues) are hiding out in suitcases, as you leave baggage claim. Also interestingly, the clear mark of the masons exists throughout the facility. It is even engraved on the granite time capsule inside the terminal.

Many people get the creeps when they walk through this airport.

All of this, of course, pales in comparison to the greeting people receive as they actually approach the Denver International Airport. At the outside entrance to the airport stands "Mustang," on Pena Boulevard.

Bluecifer

Image Credit: notmytribe.com http://www.bing.com/images/search?&q=bluecifer+denver+airport&qft=+filterui:license=L2_L3_L4_L5_L6_L7&FORM=R5IR37#a

Mustang is a 32-foot high, blue fiberglass sculpture. He rears, fiercely, on his hind legs. Mustang also has large, red-glowing eyes and his ribs appear emaciated---with veins bulging out from the side. Nearly anyone who looks at the raging horse is reminded of Satan.

The artist who created Mustang, Luis Jimenez, actually died while creating him. Mustang's head fell on him, severing his artery. On top of all of this, Mustang is nicknamed "Bluecifer," by the public.

On August 23rd at 02:40PM, twenty-seven levels below the surface of Denver International Airport, President John Diaz and General Charles Jacobs arrive at the RAPTOR terminal. The Secret Service team exits the RAPTOR and the President and General emerge. They are greeted by Admiral Steven Whistler and General Shane Maxwell.

Admiral Steven Whistler salutes the President and General Jacobs.

"Good afternoon, Mr. President…General."

The President returns the salute and tips his head to General Maxwell, who also is saluting.

"Good afternoon, Admiral...General."
General Jacobs returns their salute.

General Maxwell says, "Good afternoon, gentlemen. This way please." General Shane Maxwell leads the group to a secure conference room. Already present and seated are Lori Sanders, her support team, along with Colt Jackson. Both were rushed back to Washington D.C. after the surveillance operation on the Hotel Marsa Malaz in Qatar was completed.

General Shane Maxwell returns Commander Colt Jackson's salute when he enters the conference room.

"Hello, Commander, Hello Lori. Good to see you are alive and well. Great work on the surveillance. The President, Admiral and both Generals have been briefed on the plans of the Islamic State you uncovered at The Pearl in Qatar."

Admiral Whistler says, "Hello, Commander, hello Ms. Sanders."

"Hello, sir. Good to see you, Mr. President."

The President replies, "I should have known you two would be involved with the surveillance operation. Commander and Ms. Sanders---outstanding work!"

Lori Sanders replies, "Thank you, sir."

"Thank you, sir. There were a lot of people involved with making this happen," Colt answers.

"I'm aware, Commander. I understand some of our investments in technology paid off handsomely in helping you men in the field."

"That's correct, sir. We have the NSA, the CIA and Admiral Whistler's work with DARPA and the Office of Naval Research to thank. Ms. Sanders led one of the design teams which created the GPS and radar algorithms used to make the mosquitos intelligent. The Intel we were able to retrieve from those nano-bots may save millions of lives, sir."

"Ms. Sanders, this I didn't know. Congratulations, your work will save millions---of this I have no doubt."

"Thank you, sir."

General Maxwell calls the meeting to order. "Ladies and gentlemen, if I may…The President is on a very tight schedule and he needs to return to Colorado Springs, ASAP, before the press corps realizes he is over 80 miles away and he misses his tee-off at the Broadmoor Hotel. Admiral, you want to start?"

"Sure, General." A Power Point presentation pops up. "On October 11, which is Yom Kippur, the Islamic State will have deployed three oil tankers out of Maracaibo, Venezuela and three cargo ships out of Amsterdam. The latter are being dispatched from LUBEROL, a Dutch company controlled by Qatar Chemical Company, QEMCO. Each ship will carry a 2-10 kiloton nuclear device, attached to a North Korean mobile missile launcher."

General Maxwell discusses the next slide. "They have cleverly masked the missiles and nuclear weapons to make them invisible to our signature-sniffing satellite capabilities. Not even the local radiation detectors will pick them up until the missiles are launched."

The Admiral looks around the room with a serious expression on his face, "Frankly, if we didn't have this Intel, we would be sunk."

General Maxwell turns to Ms. Sanders, "Would you like to give the estimate on damage if the Intel was not uncovered? It seems only fitting, since you were one of the keys to making those mosquitos actually work."

"Yes, sir, General." Lori continues, speaking to the group. "The damage would be nothing short of catastrophic."

The President asks, "How many deaths would there be, initially, after launch and detonation, Ms. Sanders?"

Lori pauses and then says, "None sir."

The President questions, "How's that?"

"Well, sir, this is a planned EMP attack, which will be coordinated with comprehensive, on-the-ground, sleeper cell attacks on our electric grid. As such, the nuclear weapons themselves will be detonated above ground via ground-to-air missiles."

The President continues his questioning. "How far above ground?

"Well, sir, that is a very good question, with a complicated answer."

"Please feel free to expound, Ms. Sanders."

"Yes sir. We believe they will use the same kinds of missiles discovered on July 15, 2013, when the 14,000-ton Chong Chon Gang North Korean flagship was boarded on suspicion of drug running, right before it entered the Panama Canal. It turns out it was in Cuba and was carrying several SA-2 missiles and two Mig-21 jet fighters, hidden among 21,000 pounds of sugar. The SA-2s have many different versions among the various countries which use them."

Lori flips up a map of countries colored in red which use the SA-2, primarily Russian, formerly Eastern Block, and Asian countries. She points to the map, "These missiles are derived from the old Soviet-era S-75 Divinas, which come in many makes and models. The North Koreans have nearly three hundred of them. The SA-2B could engage targets at an altitude of up to 30,000 meters, or about 19 miles high. It has a range of up to 34 kilometers, or 21 miles."

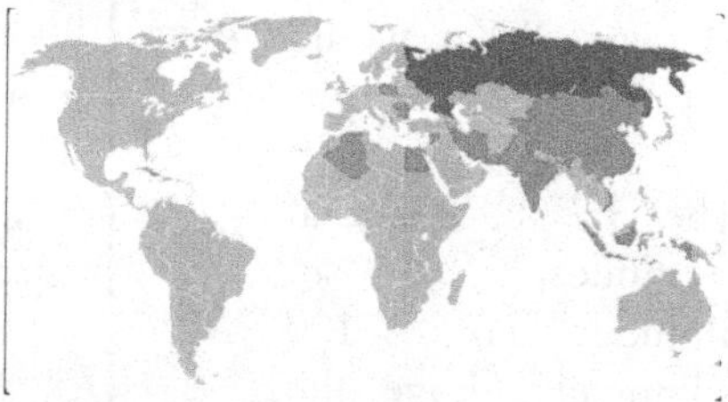

S-75 Operators

Public domain image.
Date 13 November 2008, BlankMap-World, Author User:TheGerm

The President goes straight to the point, "It doesn't sound too complicated to me, looks like they can pack a real punch on the backs of those SA-2s!"

Ms. Sanders replies, "Yes, sir, they can." She looks at General Charles Jacobs. "General, can you please fill the President in from the perspective of Space Command and NORAD?"

"I'm happy to, Ms. Sanders." The General looks from Ms. Sanders to the President. "Sir, the problem is even further complicated because of the

Taepodong-2, renamed the Unha-3. This is missile technology the North Koreans developed for their Unha space vehicle. A three-stage rocket has a range of 4,500 kilometers, or 2,782 miles. They could rain havoc on us with this propulsion system, if it was built inside a vessel with its support structure. The Unha-3 is a three-stage rocket which oxidizes the fuel, hydrazine, with red fuming nitric acid. The three stages, together, burn about six minutes, giving it enough boost to reach outer space. In fact, they have used the rocket system to successfully put their "weather" satellite, "Bright Star," into polar orbit. Bright Star flies right over the United States and both poles at 500 kilometers---about 360 miles up. This proven capability could easily launch another satellite, again in polar orbit, equipped with a nuclear weapon. All they would need to do is fire it from the satellite, or even a ground, or sea-based platform, and it could reach anywhere over the United States. If they explode it 100 kilometers up in the sky, we would have a nation-wide EMP event which could cripple us."

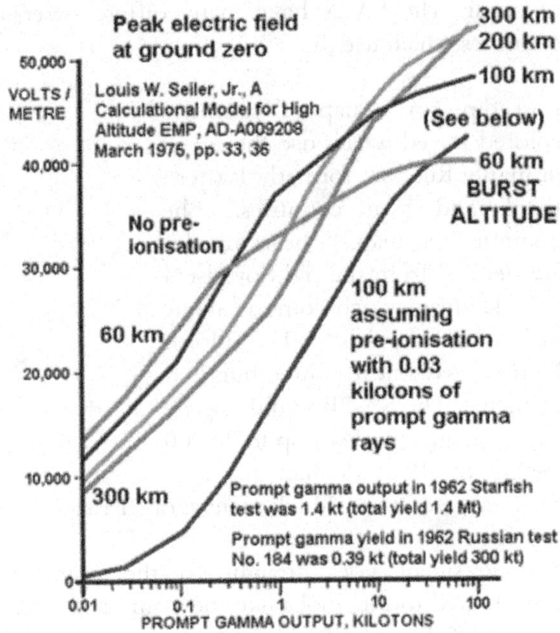

Peak Electric Fields
Ground Zero Nuclear EMP

Work product of U.S. Government. Public domain image. Composed of calculations on pages 33 and 36 of Louis W. Seiler, Jr., "A Calculational Model for High Altitude EMP", AIR FORCE INST. OF TECH., WRIGHT-PATTERSON A.F.B., report number AD-A009208, March 1975, http://stinet.dtic.mil/cgi-bin/GetTRDoc?AD=A009208&Location=U2&doc=GetTRDoc.pd

President John Diaz is definitely concerned, "I see. What would happen?"

The General continues, "At an altitude of 100 kilometers above the earth's surface, a relatively small nuclear weapon explosion, of only 10 kilotons, would cause over 50,000 volts of electricity PER METER to hit earth's surface in the United States." The General pauses to let this sink in. "This would wipe out Silicon Valley, every bank, every ATM, every television broadcaster and every electric utility. Our complete infrastructure as a country would be gone, for years. To give you an idea of the problem of recovery, to replace just ONE key transformer at the big power stations takes ONE YEAR TO MANUFACTURE. They are built in Germany and China, serially produced. People will be at each other's throats on your watch, sir. It will be ugly, very early. Ms. Sanders is that enough, or should I go on?" He then turns it back to Ms. Sanders.

"Thank you, sir. I think you've described the potential for disaster, perfectly. Mr. President, this is what we call a High-Altitude EMP event. It is further complicated because we don't have as many missile defenses pointing toward the South Pole---we don't have enemies down there. The North Koreans could be using the Muslims to send up the smaller weapons and wreak havoc on the coasts, given the SA-2s limited range and ceiling. In the middle of that chaos, they could easily launch a 10-kiloton nuke from the satellite, when it is positioned to the south of the United States in its polar orbit."

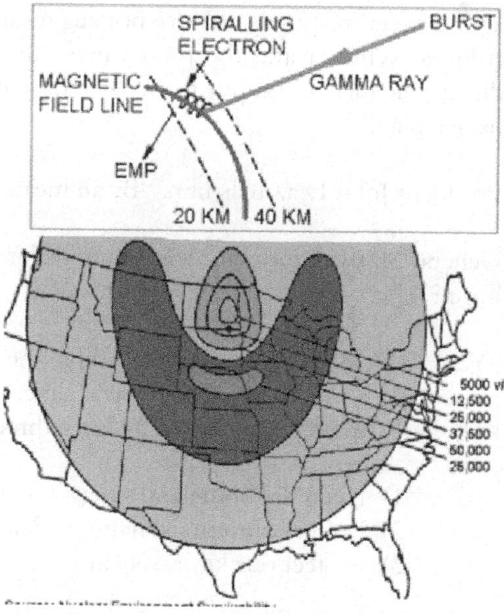

Maximum EMP Damage South Of Detonation

Earth's downward tilt and gravity cause electrons to deflect sideways, causing a massive blast area when a nuclear weapon is detonated at high altitudes. Source:http://en.wikipedia.org/wiki/Nuclear_electromagnetic_pulse
Work product of U.S. Government. Public domain image.

The President responds, "I have the picture. A sneak attack on our Southern flank--while we are trying to recover from attacks on every coast---East Coast, West Coast and the Gulf, would destroy us."

The General affirms the President's understanding, "Mr. President. You have summed up the dilemma we are about to be placed in."

The President asks, "What can be done about this? I mean, we know about the Venezuelan vessels and we know about QEMCO and LUBEROL, why not just attack them en route? By the way, what if they use an even bigger weapon in an EMP attack?"

The General answers, "Sir, if we attack the vessels, we'll have blown our asset inside the Islamic State. At all costs, we do not and should not do this. And if they use a large nuke for an EMP attack at high altitude, it is somewhat self-defeating. Larger, thermonuclear weapons can pre-ionize the air, which then becomes conductive, and hence rapidly shorts out the Compton currents generated at the fusion stage. Thus, a small, pure fission weapon with a thin exterior casing, is way better at causing EMP damage than most multi-megaton nukes."

Heatedly, the President thunders, "Regardless, I'm not going to sit around and be the first President to have all three coasts nuked via a 20-mile high EMP attack, while simultaneously suffering from a HEMP event where everyone in America will go back to the stone age because of a 60-mile high nuclear explosion!"

The General responds, "We're not suggesting this, sir. Together, between Admiral Whistler and the Joint Chiefs, the NSA and CIA, as well as the direct guidance and input from General Jacobs and Space Command, we have a plan."

President John Diaz tells him, "By all means, let's hear it!"

General Maxwell turns to Admiral Whistler. "Admiral? Would you like to begin?"

"Yes, General." The Admiral flips the button on his Power Point presentation and shows an overview slide. "We have a delicate balance to keep here, sir. As you can see we have three main objectives:

1) Keep the homeland safe;
2) Protect our critical national defense secrets; and
3) Protect our key assets in the field.

You can see, immediately, the problems. First, if we attack the ships, the Islamic State will know there was a plant inside their organization. Croc Hunter, our prized asset, our spy in the Islamic State, will be the most likely suspect, as they have not grown up with him."

General Charles Jacobs immediately asks, "Who is Croc Hunter?"

The President echoes the sentiment, "Yes, who is Croc Hunter?"

General Maxwell interjects, politely, but firmly responds, "Per Presidential Directive and operational security directives, this name is known to only a handful of people, and only on a need-to-know basis, of course, with all due respect, Mr. President and General."

The President says, "You're right, General. Let's keep moving."

Admiral Steven Whistler continues his briefing, "I guarantee you, if we jump the ships, Croc Hunter will be the star of the next You Tube video---with a scimitar hanging over his head. Secondly, we have space-based assets which can destroy the North Korean satellites. Our satellite-killer technologies are designed to take out enemies in the event of a global war against the Russians and the Chinese. If we tip our hands to these capabilities, we will lose the advantage of surprise against them and they, undoubtedly, would develop countermeasures. Thirdly, we have another layer of defenses. The CIA and NSA brilliantly stirred up and inspired a conspiracy theory. Conspiracy theorists have labeled it "Chemtrails."

The President is somewhat knowledgeable on the subject, "I'm well aware of the press coverage and questions about Chemtrails. Most of the coverage indicates the trails are designed to prevent global warming, but the Air Force has never confirmed, or denied, the existence of any program creating these strange cloud formations."

The Admiral, looking at the President, says, "You are correct, sir. But this is all a cover story. Yes, there are certain benefits which take place by seeding the sky with aluminum oxide. The clouds hang around longer and reflect the sunlight back into the atmosphere. This helps reduce the temperature here on earth due to fossil fuel emissions, carbon dioxide and other man-made activities. But the real secret, sir, lies with the passing of a famous nuclear scientist by the name of Edward Teller."

"I remember him," says the President. "He was the scientist who said a meteor could hit the earth and we had to build weapons to protect us from this kind of event. That's him, right?"

"Yes, sir. He was trying to give the press the story line we needed to develop our laser-based, anti-ballistic missile weapons. You have been briefed on Boeing's High Energy Laser Mobile Weapon System. It can knock down rockets, artillery, mortar, cruise missiles and UAVs.

Directed-energy technology can be used to protect troops on the battlefield AND THE HOMELAND. From earth, these systems still don't have the range to reach space, or extremely high altitudes, because the laser beam loses its strength, in proportion to the distance it travels. A 10kW system can travel about 1.5 kilometers and a 100kW system can travel nearly ten times that amount."

The President follows along, "So, the stories about meteors were just a ploy to build lasers which can protect the battlefield and homeland from an earth-based enemy?"

The Admiral looks over at General Jacobs. "General Jacobs, we're quickly getting onto your turf, do you want to explain how "Star Wars" evolved?"

"Sure Admiral. Mr. President, it was, at first. In fact Ronald Reagan's original "Star Wars" program was used to begin building these capabilities. But, as the science became reality, we learned we could easily attach the systems onto our satellites in space. There, with no atmosphere, the lasers work like magic from incredible distances. In a word, sir, if the Russians, or Chinese, even think about firing up one of their space-based nukes, we can turn their entire satellite and nuclear-warhead payloads into floating stones."

President John Diaz is bristling, "I'll be damned. Why wasn't I briefed on this, before?"

General Charles Jacobs responds, "National security protocol, sir. These capabilities are also on a need-to-know basis. They are one of the most highly-guarded secrets in the U.S. Defense Department."

The President asks, "Does this mean we can just zap North Korea's satellites and turn them into floating rocks?"

"We certainly could, sir. This would eliminate the HEMP threat from 100 kilometers up in the sky."

"Well, then, let's do that!" cries out the President.

Admiral Whistler jumps in. "Well, if we do this, then we've shown our cards to the Russians and the Chinese. This would surely eliminate this doomsday-defense mechanism we have in place, should the world sail into a dark storm."

General Jacobs continues, "But sir, there is one other secret, which might help us thread the eye of the needle and keep our most secret defenses under wraps."

The President is getting frustrated and is somewhat sarcastic in his response, "And what is this secret, a radiation vacuum cleaner, which immediately turns radiation into an inert element, eliminating its threat after we let them nuke us?"

General Jacobs responds, "No, sir, but I assure you, the Department of Defense is working on this, also! The big secret, however, lies in the Chemtrails, sir."

The President remarks, "How's that?"

The General explains, "Going back to Edward Teller, he also talked, thirty years ago, about creating an electronic mesh around America and its vital interests."

The General flicks on a button and up pops the Southeastern United States with an enormous amount of Chemtrail activity appearing on the slide. "Here you can see the Southeastern portion of the United States. See how all of our vital interests from Huntsville, Alabama, to Norfolk, Virginia, to the military bases in Florida, and the critical infrastructures are covered by smoke?"

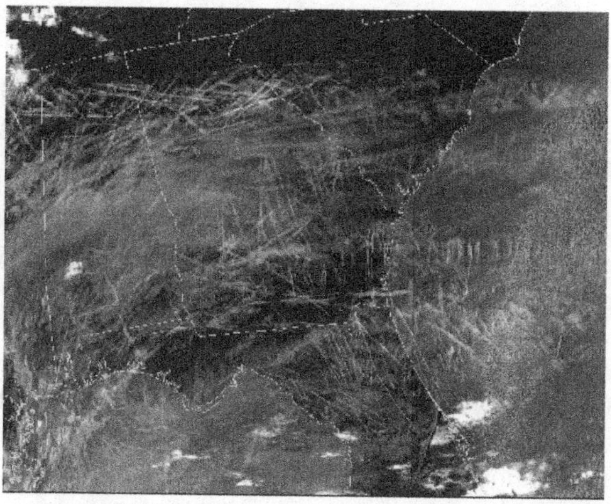

NASA Satellite Southeastern United States

U.S. Government Work Product
Public Domain

President Diaz says, "Yes, it looks odd when you see a satellite snap shot of it---that is for sure."

"Those are Chemtrails sir, except they really aren't chemical." The General pauses for effect "…They are metal. They are seeding and creating cloud cover with ionized aluminum oxide---which is also an outstanding conductor of electricity. In April of 1990, The Hughes Aircraft Company filed a for an invention under Patent #5003186, titled, Stratospheric Welsbach seeding for reduction of global warming. Quoting verbatim sir…" General Jacobs reads from the patent application which now appears on the Power Point.

> "The method comprises the step of seeding the greenhouse gas layer with a quantity of tiny particles, which include Welsbach materials. Welsbach materials are oxides of metal, including aluminum oxide. The greenhouse layer extends between seven and thirteen kilometers (4.2 to 8 miles) above the earth's surface. The seeding of the stratosphere occurs within this layer. The particles suspended in the stratosphere as a result of the seeding provide a mechanism for converting the blackbody radiation emitted by the earth at near infrared wavelengths into radiation in the visible and far infrared wavelength so that this heat energy may be reradiated out into space, thereby reducing the global warming due to the greenhouse effect.*"

Admiral Whistler expounds, "You see, sir, Edward Teller was not only a genius, but he was a mastermind at deception. While helping the earth solve a fictional greenhouse problem, Dr. Teller saw something far more evil on the horizon. This was our growing dependence on the semiconductor and the electric grid. Dr. Teller knew, because of his experience in building nuclear weapons at Los Alamos and Lawrence Livermore National Laboratories, the threat of an EMP attack was one of the most severe threats our nation faced. But, he also knew how to surreptitiously SOLVE IT!"

The President looks astonished. "HE SOLVED IT? PLEASE TELL ME MORE!"

The Admiral continues, "Dr. Teller and his colleagues designed a number of mixtures to be used in the Chemtrail seeding program. One of them was a mixture which created a sky which became a virtual electrical-conducting sheet. Dr. Teller called the sky after it was seeded, a "metallic net" in the stratosphere. Aluminum oxide, mixed with an aerosol of

* #5003186, "Stratospheric Welsbach for reduction of global warming." Application # US 07/513,145. Publication date, Mar 26, 1991. Inventors David B. Chang, I-Fu Shih, original assignee, Hughes Aircraft Company. USPTO.

barium salt, provides an amazing conducting path. It is outstanding for communications, BUT it also creates a natural defensive barrier to the highly charged, intense atoms of an EMP attack! As the General already answered your question on why a large, thermonuclear weapon is not ideal for an EMP attack..."

President John Diaz interrupts with excitement, "I get it! The metallic mesh of the Chemtrail between the greenhouse gas layers between 4-7 miles above the earth will pre-ionize the air. It then becomes conductive, and will short out the Compton currents and high-energy voltage generated by the fusion, or fission, stages of an EMP, or HEMP attack!"

General Jacobs in a congratulatory manner says, "That's it, Mr. President! We call this breakthrough defense mechanism, 'Angel Shield.' It is an integral part of America's top secret defense against objects in space, as well as the detection, validation, and warning of any kind of airborne, or seaborne, attack against North America. Angel Shield can identify friend, or foe, whether be it an aircraft, missile, space vehicle, or any maritime threat which might exist. The shield was originally built and designed to withstand a surprise offshore, or space-based nuclear attack, from Russia, China, Pakistan, North Korea, India, Israel, South Africa, France, Great Britain, or any other country on earth which might own weapons of mass destruction. The knowledge of its existence is one of the best kept secrets in the United States."

Not even the President knows the full extent of Angel Shield's capabilities. The components creating Angel Shield are highly compartmentalized. Only the Joint Chiefs of Staff, the head of the CIA and the head of the NSA and the Commander of NORAD know of its integrated capabilities.

A big decision will need to be made in the near future as to whether, or not, to alert the world to the fact America has an invisible, protective, electro-metallic barrier around it, called the "Angel Shield."

General Jacobs continues, "We believe with an aggressive use of Chemtrailing across the United States, we are impervious to serious attack from an EMP. We will likely suffer some damage, but the damage will be containable and manageable. We can definitely ensure the power grid is safe and all the major cities are protected, along with our national defense interests."

The President starts worrying about other parts of the nation, "What about the small towns?"

General Jacobs replies, "They will also fall under the umbrella of Angel Shield, sir. We'll have the entire country covered, with a virtual Faraday cage. Faraday cages prevent electrical devices from being zapped during a storm, or during an EMP attack. They are simply metal mesh placed around an electronic device you need to protect, with a protective layer in between the object and the mesh. Our atmosphere provides us the rest of the protection. Any ambient energy remaining should cause only minimal damage."

Admiral Whistler adds, "In the meantime, the Islamic State can be put against the ropes and bombed into oblivion. We can also arrest Emir Sheikh Muhammad Al Therani for war crimes and a conspiracy against humanity."

Colt Jackson, stirs at the far end of the table. He addresses General Jacobs, "Sir, if I may, I have a question?"

"Commander, please go ahead."

Colt replies, "Not to be Johnny Raincloud, but what if "Angel Shield" doesn't work?"

Everyone stares at each other around the room, in silence.

Colt continues, "I mean, aren't we risking our entire infrastructure and country on an untested "metallic net." I'm no techie, sir, but this seems like an awfully big risk to me. I don't like it. We're going to sit back and let a half-dozen nukes go off and do nothing about it, except play possum?"

General Jacobs is looking displeased with Commander Colt Jackson. "Commander, you are one of the most respected people in our entire military. Of this, there is no doubt, but Commander---this subject IS a bit above your pay grade. NORAD, USNORTHCOM, Los Alamos, Lawrence Livermore, Oak Ridge and the entire U.S. Space Command have tested this via simulation on our most advanced supercomputers. This was done on the Titan system, at Oak Ridge National Laboratory in Tennessee, on the CRAY XK7. It scales to 500,000 scalar processors and has 50 petaflops of hybrid peak performance. A single petaflop is ONE THOUSAND MILLION floating point operations per second! We are

99.99% confident in these simulations! We even performed an actual test of the ability of the metallic net to shield earth by bombarding a metallically-treated, sprayed Chemtrail area, with high-energy ions. The high-energy ions were all short circuited, with 100% effectiveness."

The President comes to the aid of General Jacobs, "Commander Jackson, prior to this briefing, I wasn't knowledgeable about the detailed specifics of how Angel Shield works. However, I did receive an overview about the science, a couple of years ago. It documented the results of the actual test performed over the South Pacific Ocean. Every single high-energy impulse was defeated and short circuited. The details on how it actually operated were, of course, left out---at least until today's meeting. Now, it all makes sense. Chemtrails---really Metrails---are the key. Angel Shield works. I like and admire your instincts to protect America, Commander, but, in this case I believe you are overly concerned. You are ignoring the facts of the test and the results of the simulations."

The President looks at General Charles Jacobs. "Great job, General."

General Charles Jacobs now displays a look of triumph over Colt Jackson. "Thank you, Mr. President."

Admiral Whistler makes a suggestion, "Mr. President. To Commander Jackson's point, I have a suggestion---it's not in the slightest way impugning the veracity of the General's claims on the metallic defense net of Angel Shield."

"What is your suggestion, Admiral?"

The Admiral continues, "The Qatar Chemical Company, QEMCO, is using its European distributor, LUBEROL, to deliver missiles and payload to the Chemrol Corporation in New Jersey, for an East Coast attack."

The President breathes a sigh, "Yes, we know this, thanks to Ms. Sanders' mosquitos, thank God."

"Exactly, Mr. President. I believe Croc Hunter could be completely protected if we went after the LUBEROL shipment while it was is in the Atlantic. ISIL's attention would be focused on a leak in Amsterdam, not out of the Islamic State leadership, itself. General Maxwell, what do you think?"

General Maxwell and Admiral Whistler are long-time friends and watch each other's back.

"Admiral, I believe you are right. The Islamic State would assume the Amsterdam connection was corrupt. They would look to chop heads there---literally. Mr. President, the NSA concurs with the Admiral's suggestion. This would at least ensure nothing would happen to the East Coast."

"Very well gentleman. I'll leave the details for you to follow up on. In the meantime, General Jacobs, spray the entire country just before the attack date set for Yom Kippur, October 11th. I think this pretty much wraps it up. General, can you take me back on the RAPTOR? Perhaps I can get that golf game in."

General Jacobs replies, "Yes, sir!"

Everyone departs the meeting room.

At 03:55PM, Colt Jackson and Lori Sanders emerge from the elevator and appear in the middle of the concourse of Denver International Airport. They are walking together and brush up against each other's shoulders. There is enormous chemistry between them.

"I don't know, Colt. I'm not sure I trust the simulation and test results, either. There is just something weird about relying on an unproven, national metallic net, to fend off a nuclear EMP attack on the homeland. At least your questioning allowed us to eliminate the threat on the East Coast."

"For me, Lori, it just doesn't pass the sniff test. Something about General Jacobs---he was overly confident in the computer simulations---or, perhaps, trying too hard to convince everyone this virtual metallic net could protect us. This is why I questioned him. It seemed suspicious to me, with so much at stake."

"I agree. What should we do?"

Colt says, "We have to follow orders, of course, but no one said we couldn't do a little digging!"

"Gotcha, Commander!" Lori smiles and winks at Colt. Colt bubbles with energy when he looks into Lori's beautiful, green eyes.

"Ms. Sanders---would you like to have dinner with me, this evening?"

"I'd love to, Commander. We're all booked at the Grand Hyatt downtown Denver on Colorado Street. I'd like to freshen up, if I could."

"I brought my swimming trunks because they have an indoor pool. I need to stretch and exercise a bit. The flight and the meeting have taken their toll."

"So, 7PM"

"Perfect, Lori. See you then."

"Oh, Commander? Remember, I'm not squeamish. How about Rocky Mountain Oysters at the Buckhorn Exchange?" Lori smiles a dazzling, seductive smile. Colt is blown away.

"You are too much, Ms. Sanders! TOO much---and way too hot to handle! I'll knock on your door at 7 sharp. Rocky Mountain Oysters, it is!"

Lori tells Colt, "I have to quickly debrief with General Maxwell. I'll see you back at the hotel." Lori is gazing at Colt. They turn closer to each other, staring intently, and embrace each other, firmly holding their bodies close together. Colt is stirring inside and needs to take off.

"Until then, Ms. Sanders, I need to take a cold swim!" They laugh and gently separate. Each goes their separate way in the airport.

CHAPTER 12

THE OLD BUCKHORN

At 07:25PM on August 23rd, Lori Sanders and Colt Jackson have taken a cab to the restaurant. Denver's Buckhorn Exchange Restaurant's fabled, red and white awning, with the elk's pointed antlers and its old, weathered and sun-bleached brick exterior, reminds anyone who ever visits of the old West.

The Buckhorn Exchange

Source: Wiki Commons http://commons.wikimedia.org/wiki/File:Buckhorn Exchange.JPG Date 30 May 2009, Author Jeffrey Beall's own work. Permission (Reusing this file) CC 3.0

Lori and Colt have never eaten here, but each of them has heard of its famous Rocky Mountain Oysters, more commonly known as bull's balls. The Buckhorn's menu has fresh game, fish, steaks and meals come with delicious veggies. This makes the restaurant a favorite of locals and tourists, alike. The Buckhorn Exchange Restaurant is only about 11 blocks to the southwest from the Grand Hyatt downtown, at the corner of Osage and 10th Street, where Colt and Lori are staying overnight. Appropriately, this old-west destination sits adjacent to the rail line, another symbol of Denver's golden, western heritage. The Buckhorn opened for business over a hundred-and-twenty years ago, in 1893.

Lori and Colt enter the restaurant. They immediately notice the red and white checkered tables are watched over, intently, by 575 pairs of dark eye balls. Hundreds of mounted, stuffed animal heads and stuffed animal bodies, from all over the world, surround diners on all sides, while they are eating. These include moose, mountain lion, bear, elk, deer, rabbit, water buffalo, bison, pheasant, badgers, wolverine, beaver, antelope and dozens of other species. Colt notices the enormous, old-west gun collection. He admires the old Colt .45s, Winchesters and a Sharp's sporting rifle dating to 1889.

"Ms. Sanders, can I show you something? See this pistol?" Colt points to the famous 'Colt .45' in the restaurant's gun collection.

"Yes, what about it?"

"I was named after this gun. The 'Colt .45'---my dad loved the name and our ancestor, Stonewall Jackson, loved the weapon. He was friends with the manufacturer, Samuel Colt."

"You are a chip off the old block, I believe, that is for sure!"

Lori walks to their table in skin tight, J-Brand Maria High Rise Skinny Jeans. She is also wearing suede, spiked heels. She is a knockout.

Colt admires her beauty and knows she is a dream. She is not wearing her glasses, either. She doesn't need them. They are only for show at the NSA and her light brown, blond-streaked hair flows invitingly over her shoulders.

Colt tries to absorb her raw beauty and quickly finds he is looking at her Vince, Shoulder-Pleated, V-Neck Stretch-Knit top. It complements her figure amazingly, as the plunging neckline turns every man's head in the restaurant. Lori tries valiantly to dress conservatively when she is at work; tonight, however, on her free time during a relaxing dinner, Lori's stretch-knit top and tight fitting J-Brand jeans reveal a vivacious figure. She is quite the sight and quite a woman.

For his part, Colt is dressed in his best casual, too. Right down to his Wrangler jeans and his black Hugo Boss, V-neck t-shirt. He bought the shirt on a whim one day at the Georgetown Park mall, downstairs from his condo in Washington D.C. It was on sale. The shirt was supposed to be loose fitting on most men, but a careful inspection of the shirt on Colt, reveals the strength of a chiseled body underneath it. Colt is wearing his favorite thing when he is on his own down time---his Lucchesse full-quill ostrich, camel-colored cowboy boots. They were a gift from his dad before he died. They wear like iron, but are as comfortable as a pair of Nike tennis shoes.

Colt remembers his dad told him he should always have a great pair of cowboy boots, to remember where his family came from.

Colt and Lori make a very striking couple, together. They sit down and look into each other's eyes.

"Lori, I'm speechless. You are breathtaking this evening. I really mean it. I can barely control myself! I love being able to dress casual---at least by modern-day standards---and be comfortable together. No matter how you dress, you are simply amazing."

Lori smiles evocatively at him and her beautiful green eyes glow in the restaurant lighting. "Thank you, Colt. I love wearing jeans! I must admit, you look pretty good, yourself. You certainly look better than Brett Favre in those Wranglers on TV!" She laughs.

Colt is getting flustered and he laughs nervously. "I don't know, I just like them because they fit me and aren't too tight. The designers these days, for most jeans, make them too tight for my body. I feel like I'm going to split the crotch every time I try to put a pair of on! Diesel's are just too tight for me. I am doomed to the original western wear, I guess, but I never had any desire to be a fashion statement."

Lori starts laughing hysterically. "YOU don't have to worry about that! Despite the fact YOU ARE a dinosaur in the clothing department, you are extremely handsome! I think every woman turned her head to look at you when you walked by."

"Very funny, but I know EVERY man turned his head looking at you. I found myself issuing each of them a cold stare as I walked by!"

Lori looks deeply into Colt's cobalt-blue eyes and smiles, "I am purely and only happy to be here with YOU. Thanks for asking me out, this evening. This place is picture perfect for us while we're in Denver. I have heard great things about the Buckhorn Exchange. The ambience is totally unique."

"Lori, after dinner, maybe we can go upstairs to the bar and have an after-dinner drink? The bar was made in made in Essen, Germany in 1857. It's 100% white oak---you know, one of those old-style saloon-hall bars. They have a whole Victorian lounge upstairs. This place actually received the State of Colorado's first liquor license and I think they have some music up there. What do you think?"

"That would be great! I wonder…what should we order??" Lori has a mischievous smile on her face and raises her eyes at Colt, flirtingly. "I mean, of course, besides Rocky Mountain Oysters!"

Colt laughs, "Well, why don't we go for it. It will be too much food, but our chances to be together like this are precious. Why not make it a night to remember?"

"Colt Jackson, I simply love how you think. Do you think we should try the rattlesnake dip and artichoke hearts, as well, for appetizers? The chipotle dip with fresh tortilla chips sounds really interesting."

"Sounds good to me. What about dinner?"

"Hmmmmm, let's see."

Colt spots the lamb and says, "I might try the Colorado lamb with scallion-topped baked potatoes and some sautéed vegetables."

"Yummy. I think I'll go for the combination dinner of the quail, duck and Cornish hen. Do you think I can try your lamb, we could share?"

"Perfect! Do you want white wine, with your bird?"

"Actually, there is a Kendall Jackson Cabernet Sauvignon on the menu. I love Kendall Jackson."

"OK, it sure goes with my lamb. I know this is pushing it to the max, but after my swim, I am voracious. Shall we put in an order for the Double Chocolate, Rocky Road Brownie---we can share it?"

Lori is looking intently into his eyes and continues smiling, flirtingly. "What girl doesn't like chocolate? Mmmmmm…YUMMY!"

They order their meal, and despite the incredible sparks and obvious romance of the evening, they go back to business for a moment.

"You know, Lori, I was thinking about what we were discussing earlier. General Jacobs seemed defensive, overly defensive, didn't he?"

"It's been bothering me, also. It may be all legit, but I need to dig into the background on the test and the simulation. I'll do this back at Ft. Meade when we return tomorrow. OK? You were right on the money to question him like you did during the meeting. I am always amazed at your ability to surf through tons of information and get right to the point."

"I think it's just an instinct I have developed which has kept me alive---at least this far in life. Many battlefield commanders have a 'spider-sense.' Maybe I have some of that, too. I don't know, but I read people extremely well and there is just something I can't put my finger on about General Charles Jacobs. Maybe it is his avoidance of eye contact at key points in time. I'm not sure, but I feel tingly all over when he is speaking."

Colt Jackson and Lori Sanders finish their delightful meal and can barely walk, they are so full. They head upstairs and enjoy the over-stuffed Victorian, cozy sofa. Colt orders a Courvoisier and Lori orders a Sambuca. They enjoy the music and hold hands, allowing the sides of their bodies to rub against each other on the love seat. They enjoy listening to the music and the evening melts away. They finally depart, because they are both on early flights in the morning.

At 11:30PM they get back to the hotel. Colt and Lori exit the taxi and walk into the entrance of the Grand Hyatt. They walk through the lobby and enter the elevator. They each punch different floors. The elevator comes to a halt and the door opens.

Lori says, "Looks like it's my stop! You going to walk me home, Commander?" Lori smiles.

Colt smiles widely, "ABSOLUTELY!"

Colt walks Lori to her room. They both look at the door and then at each other. They embrace---then wildly kiss each other in a fit of passion. Lori's breast swell up against Colt and Colt becomes beyond excited. He is about to reach the point of no return, when they both gently, but grudgingly pull back. They know as Christians they should not have sex before marriage. They look deeply and longingly into each other's eyes.

"Colt, I don't know if I can control these feelings I am developing for you. They are very strong."

"Lori Sanders, my head is about to explode. I think I will go crazy---even if I walk away right now for the evening. Our bond is very strong. We just work, don't we?"

Lori's eyes look intently and lovingly at Colt. "Yes, we do." She collapses one more time in his arms. They kiss deeply for minutes. Colt cannot control his manliness and he grows against the soft, but firm and warm body of Lori Sanders. Colt's animal instincts are nearly primordial, as he feels Lori's firm breasts rub against his aching body.

"Colt, I am getting so hot." She kisses him again and he holds her passionately in his arms.

Colt, very reluctantly, pulls his lips away from hers---just enough to be able to look deeply into her emerald-green eyes with his blue-eyed stare. "I know Lori, me too, in more ways than one. We either have to break the glass ceiling, or we must somehow walk away, for now, and take cold showers, very long, cold showers. I truly think I will explode with desire for you. I want us to do this right---it could be for all the marbles, you know?"

Lori looks longingly and lovingly into Colt's eyes. "I do, Colt, I do." Lori, realizing those are the words spoken when getting married, quickly makes an evasive maneuver with another mischievous remark. "I mean, a girl can't be too careful now, can she?"

Colt feels somewhat embarrassed. "I'm sorry I am getting too excited. It's very difficult to control. I think we should say goodnight, before we do something we will regret. I want us to always have this, Lori, this, right here. This longing, this...this...love."

Lori Sanders eyes are sparkling. She smiles widely and inserts her electronic door key into the lock. Her door opens and they both look at each other. Her bed is in full view. They glance at it, and then look at each other. Colt leans into her and kisses her, deeply and powerfully. Their madness in passion is overwhelming. Lori pushes Colt off of her, firmly holding his powerful shoulders, but not letting go. She stares deeply into his eyes. "Colt, here I am, all for you. But I will only have you be all for me. When we get to that point, we won't have to question anything. We will know what to do."

She leans up and kisses him quickly, but sweetly on the lips. "But, right now, mister, you better scram. I don't think I can even catch my breath any longer and I am running out of will power. Thank you for a simply beautiful evening."

Colt leans down and gently kisses her, quickly, one last time. "Good night, Lori. Thank YOU for a beautiful evening." He turns and walks down the hallway to the elevator. He hits the "Up" button to get to his floor.

Lori closes the door behind her, leans against it and breathes deeply, trying to literally catch her breath. She says to herself, "*OH MY, what am I going to do about this man, he takes my breath away.*"

CHAPTER 13

LUCKYLEFTY.NET

A month later at NSA headquarters, in Ft. Meade, Maryland, it is September 23rd at 03:30PM. Lori Sanders is working on her computer in her office. She is reviewing satellite surveillance she had placed on General Charles Jacobs, done with General Shane Maxwell's approval. She had tasked the servers with tracking the General's whereabouts, immediately after her return to Washington D.C.---and her miraculous evening with Colt Jackson. She hasn't seen him since. Colt was whisked away to prepare for the upcoming Islamic State attack. The agency assigned it a code name, which stuck. Operation: Black Storm.

Lori picks up her interior office phone. "Kristi? Please come in here and bring Brian with you."

Kristi Kruger says, "Yes, ma'am. Right away."

Kristi Kruger and Brian Pierce immediately appear in Lori Sander's doorway.

"Brian and Kristi, I need you to do something for me. It is urgent. It is a national priority and it is code red. I need to know who, what, when and where General Charles Jacobs is seeing and doing in Silicon Valley tomorrow. Retrain the satellites on and over Moffett Field, where he is scheduled to fly. You are to track him with Keyhole and monitor all his discussions. I mean EVERYTHING. You two are to trade off, every three hours in the surveillance. I want a direct report from either of you, immediately, if you sense anything which seems unusual, or suspicious. Do you understand?"

Kristi and Brian both say, "Yes, ma'am. We're on it."

On September 24th, inside Kristi Kruger and Brian Pierce's computer pod, they are both tracking the vehicle of General Charles Jacobs. They have spotted nothing unusual, as of yet. There could be a million reasons why the General has to visit Moffett Field. Moffett field is the home to NASA Ames, a major center for spaceflight and information technology. Ames provides leadership in advanced thermal astrobiology and satellite technology. Supercomputing simulations and experiments are also carried out there.

Brian observes, "Nothing really to see at NASA Ames, Kristi. The business there, in all probability, is 100% legitimate. They are leaders in astrobiology, satellite technology and supercomputing. They can run simulations right at NASA Ames and hook into the Titan in Oak Ridge Tennessee by NGA---Next Generation Access, at speeds hundreds of times faster than today's Internet."

"I agree, Brian. Nothing to see there, it would seem. But, where are they going? It looks like they are heading into Palo Alto. I love The Fish Market there on El Camino Real---the Mesquite-grilled fish is the best in the Bay Area. It is to die for."

Brian looks over to Kristi. He is always trying to get her to go out with him, "Maybe we can go sometime?"

Kristi shoots him a dirty look and rolls her eyes a bit. "Nice try, Brian. Work and play---can't let them come together, you know?"

Brian looks intently at the screen. "Hey, check it out. His vehicle just went by The Fish Market and he's turning onto Olive Avenue---he's near HP's headquarters and he just passed Mike's Bikes. It looks like he's heading to Park Boulevard. Yup, that's where he's going all right. He's pulling into 3045 Park Boulevard."

Kristi's fingertips are flying over the keyboard. She is searching Park Boulevard businesses. "That's a body shop. Is his vehicle damaged?"

Brian looks at the Keyhole satellite real-time video feed. It can read a newspaper from space. "No, not at all. It looks like a brand new, bullet-proof, black Escalade. Not a scratch on it."

Kristi is looking at Brian quizzically. "What do you think General Jacobs is doing at a body shop in the middle of Silicon Valley, when he doesn't need repairs?"

Brian looks seriously at Kristi, "I'm calling Lori. I think this qualifies as unusual, don't you?"

"Absolutely!"

Brian calls Lori Sanders and speaks with her over the phone. "Yes, ma'am, that's what we're saying. No damage to the vehicle and he's going into a body shop building now."

"Brian, put me on speaker…Kristi, Brian, I want you to switch over to the infrared and tap into the audio surveillance capabilities of the PRISM system. We need ears inside. In the meantime, find out everything you can about this 'body shop' business."

They both reply, as one, "Yes, ma'am. We're on it!"

They begin searching business records, lease agreements, phone records and cell phone logs from the building. They uncover a myriad of communications to a cell phone in Colorado Springs, but there is no trace of ownership record for the number. It is as if it was wiped off of the databases.

"Look at all the calls to Colorado Springs, Brian, exactly where General Jacobs is based! Did you find anything on the leases?"

"I just found a name of the sole proprietor. His name is Lucius Cifer. I also came across a contract with AT&T for a dozen, high-speed T-4 lines. That's a lot of bandwidth! Just one T4 trunk gives a DS-4 signal level with the equivalent of 168 T1 lines!! Each T4 can support over 4,000 voice channels and provides a near light speed 274.76 million bits of information per second."

"Phewwww...I don't think they are repairing cars, do you?"

"No way! They're undoubtedly using the Internet. Let's get PRISM on this and then give Lori the report. This should be very interesting."

Kristi Kruger and Brian Pierce analyze the information dump coming at them on their computers.

Kristi speaks first, "I don't believe it. It's a social media site called luckylefty.net."

They both rush to Lori Sander's office and brief her.

Lori states questioningly, "Lucky Lefty Dot Net?"

Kristi says, "Yes, ma'am. That's it."

"Now, why in the world would General Jacobs be messing around with an Internet operation called "luckylefty.net"?"

Brian states, "That's the $64 million question. It is only about a fifteen minute drive from Moffett Field. I went back and looked. The General has been going to Moffett Field a lot in the last six months."

"Who did you say signed the lease, Brian?"

Brian looks at his notes. "It was signed about six months ago by a person named 'Lefty Cifer.'"

Lori raises her eyebrow. "Lefty Cifer? What kind of name is that? Run it through the database. I want photographs, history, background, university and any correlations between Mr. Cifer and General Jacobs. Bring me the report in twenty-four hours."

Kristi Kruger & Brian Pierce say in unison, once again, "Yes, ma'am. We're on it!"

"Kristi, Brian? Great work." Lori Sanders smiles at them.

Kristi Kruger and Brian Pierce head out the door and tip their heads to her.

At 04:12PM, Pacific Time and General Jacobs enters the secure building with an electronic key.

www.luckylefty.net is a very busy place, with servers and software programmers all over the building. It has a sprawling main floor, with an open-architecture. The General heads up to the large office, which sits over the bullpen, below.

This is the office of Lucius Cifer, also known as Lefty Cifer, who used to be known as Lucky Crockett. He has morphed himself into a new body, but similarly to Lucky Crockett, he has strong, penetrating dark-brown eyes, has charismatic charm and a smile which flashes on, at his whim. This is the man who vanished after the failed attack on the financial system of the western world. This is Lucifer incarnate, Satan's foil. Lucifer---the being that has now reincarnated himself into the body of Lucius Cifer.

Lefty Cifer is an emerging Silicon Valley titan and his new social media company has taken the world by storm. Everything is FREE. Movies and music, as well as all forms of other media, are all FREE. No one can figure out how he has paid for everything, but one thing is certain, he has crushed Facebook and Google in terms of collecting advertising revenues.

The rumors floating around are that he was connected to the banking families and the media companies. He offered them all a deal they couldn't refuse, supposedly, but the terms have never been disclosed. In six short months, luckylefty.net has collected 300 million Internet users. It is already about a quarter the size of Facebook. Luckylefty's growth is exploding.

Lefty Cifer greets the General. "Hello, General! Great to see you, my friend. How is it all going?" Lefty warmly greets the General with a big hug and pat on the back.

General Charles Jacobs replies, "It's going well, Lefty. Can we go into the private office?"

When the General arrived in uniform, all eyes from the software floor were on him.

Lefty warmly replies, "Of course!" Lefty places his palm over a panel and looks into a retina scanner. The wall slides open. The men enter the secret domain of Lefty Cifer.

The General always marvels at the interior of this room. The rich, deep, blood-red velvet wall paper, with black modern, paisley accents, is set against dark mahogany paneling. It is stunning and evocatively attractive.

The General always felt like the wall paper was alive when he walked through this room. The lair of Lefty Cifer, at ground level, is a magnificent living space. Within these walls, the General has always seen beautiful women running around. Some of them, today, are naked. They seem completely at ease and comfortable with a stranger seeing them this way.

"*They should be naked, of course,*" the General thinks. This being, Lefty, is our Master, after all. He should get anything he wants. It was Lucifer, the being of light, who found him at NORAD and NORTHCOM.

Six months ago, Lefty met him in Colorado Springs and revealed himself, after it became known General Jacobs was an atheist. Lefty knew it was a bit of a chance, but he knew the General's ambitious heart. For the General's complete loyalty, to Lucifer first and everything else second, General Jacobs was going to become the next Chairman of the Joint Chiefs of Staff. This meant the General would skip over everyone else. It would send shock waves through the military and catapult him to the forefront of the world's military elite.

His entrance to and chairmanship of the Joint Chiefs of Staff was only months away, after the country was sent into chaos by the Islamic State attack. The General became the loyal soldier of Lucifer when they completed this bargain.

Now the General is alone with Lefty, he can drop the act.

"MASTER! It is great to see you!" He bends down on one knee and kisses the red ruby ring on Lucifer's finger."

"It is great to see you too, my son! Please, get up. Is our business urgent, or do you have time to have your thirst quenched and be serviced by my beautiful family of wonder women?"

Lefty looks at him with an evil, empowering stare and a broad smile across his face. He knows the General likes rough sex with women and these women are ordered to do anything the General asks.

The General replies, "I'm afraid it is rather urgent. We are coming up on October 11, Yom Kippur for the Jews, and the day of the attack. I need to go over the last minute details and determine any changes we need to make."

"You're no fun today, General!" Lefty rolls his eyes from the General, over to a trio of stunning beauties. Lefty is eyeballing, voraciously, all at once, a gorgeous red-head, an amazing brunette and a buxom blond woman. All of them are just waiting to service them. Lefty says, with some feigned disappointment, "OK, you lovely ladies can leave, but stay close by! I'll be calling you over, soon!"

The women giggle and scurry out of site. They know Lucifer needs his privacy.

General Charles Jacobs reviews a few things with Lefty. "We are all set, Master. We had a wrinkle from a real pain-in-the-ass during the meeting in Denver. He talked the President into sinking the container vessels coming out of Amsterdam, which will be carrying the nukes for the EMP attack on the East Coast. This foiled our plan to use the "metallic net" on the East Coast, as the primary defensive tactic."

Lefty looks seriously at the General, "Who was the pain-in-the-ass?"

"A guy by the name of Colt Jackson…" Lefty Cifer's eyes immediately light up when he hears the name. "…He's a U.S. Marine Colonel and a Commander in the U.S. Navy SEALs. A real rah-rah guy. Do you know who he is, Master?"

Lefty gets an evil and dark look on his face, "Let's just say we've met in a previous life. He can be a real pain-in-the-ass. That is for sure. Keep an eye on him, that's an order."

"We will, sir. Anyway, nationwide, as you have directed, the metallic net will fail, despite my assurances on the simulations and the Pacific Ocean tests I gave to the group in Denver. This is because, as you've ordered, we will replace the aluminum oxide with a stable aerosol. The aerosol will not short circuit the ionized atoms from the EMP blast. The atoms will pass right through the Chemtrail, like a knife through hot butter. The highly-charged particles will deliver a whopping 50,000 volts of electricity PER METER, throughout the United States. The only place which will be sprayed with the aluminum-oxide will be over this area of California, down to around Santa Barbara---protecting your operations here and its links to Asia, the Middle East and Europe.

Lefty says, "This means we can expect destruction over LA, Texas, and the Gulf-coast states from the Venezuelan tankers."

The General confirms, "Yes sir. We should expect some serious damage also extending out of Charleston north. It could hurt D.C. Do you want me to make the space-based laser system fail? Then, the North Korean satellite could fire its weapon and create a HEMP. This would cover the entire country. I will do it, Master, just give the word."

Lefty replies, "I'm going to need you to survive all this, when the government tries to resurrect itself. You will be able to declare your metallic net actually works, because you will have saved San Francisco and the entire Bay Area. It will be unaffected because of Metrail, the metallic-aerosol net you will have laid down. I have already set up some fall guys in the Chemtrail squadrons who will get caught after they fill up the planes with the stable aerosol. It will look like Chinese espionage and you will be clean as a whistle."

"Your wish is my command, Master."

"Good, it's nearly settled then. All we have to decide is whether, or not, to tip off the Russians and Chinese as to the advanced, space-based laser weapon system, if you take out the North Korean satellite. I think to keep them at bay we may need to proceed with the plan to take out the North Korean satellite, if it threatens to strike. It will make you look good. Remember, our target for true takeover is a little over three years from now, following the election."

"What will be your plan, Master?"

"By then, we will have internationally taken Luckylefty.net public. At that time, I will then announce my candidacy for office, after I grease the skids of the Democratic election machine. I am targeting Las Vegas for the National Committee's big fund raiser. It is two years before the general election, in January."

"Excellent, Master."

Lefty continues, "The country will be short of money and the oil and gas industry in Texas, Louisiana and the Gulf, will be crippled, for at least a year. Hopefully, the military bases down there are not fully hardened. To the extent they are not hardened enough against an aggressive EMP attack, they will also be out of commission for a long time. I'm counting on LA and the Southeast to be decimated."

"In the meantime, Master, your plan is the same? You will short the markets in the first week of October and then use the profits from the crash around the world to gobble up assets on the cheap?"

"Indeed. But, remember, General. It isn't only about assets. It is also about acquiring the people who will help us take over the country. This is the real key."

"I understand, Master. I know you have already reconnected with your banking family heirs. They are eager to short the market again after they lost their shirts during the Great D'Apocalypse attack."

"Listen carefully, General. That Commander---Colt Jackson, he is a real wild card. You make sure he is ordered to STAND down. I don't want him screwing things up, again."

General Charles Jacobs bows to Lefty, "I will, sir." He looks around at the deep texture of the walls of the lair. "I wish I had time to stay, Master. Your hospitality is always the very finest. It is an honor to be in your presence and share your women."

Lefty Cifer sweeps his hand towards the women, who have somehow sensed the meeting is about over. They hang in their negligee, bulging out in perfect form. Lefty smiles a bedeviling smile at the General, "Are you SURE you don't have time to stay?"

"I can't this time, Master. I have to watch my time and get right back to Moffett Field. The plane is taking me back to Colorado Springs in about an hour."

Lefty Cifer looks at him intently. "That's too bad General. Eva was asking about you. She LIKED you, that's for sure."

"Please send Miss Eva my regards, Master. I look forward to seeing her again, someday soon, I hope."

"Very soon, General! Very soon! Perhaps we will pay you a visit. Would you like that?"

"I would love that, Master. Mi casa es tu casa, you know that. We can also take over the Broadmoor Hotel any time you want. I can issue a Presidential preference order at any time. Just let me know."

"I like the sound of that, General. Good hunting, my friend. And good luck!"

"Good bye, my great Master, my Lucifer!"

The General is escorted out of the lair by two beautiful, scantily dressed women who appeared out of nowhere. The wall slides open and closes behind the General.

It is now 04:45PM Pacific Time on September 24th. The General shakes his head and goes down the stairs of the offices of www.luckylefty.net. He passes through the bullpen and heads out of the building.

Back at NSA headquarters at Kristi Kruger and Brian Pierce's computer pod it is 07:45PM Eastern Time. Lori Sanders is looking over their shoulders at the computer screen and says, "Did you get any audio? Anything at all?"

Kristi responds, "We could pick up everything in Lefty Cifer's office, but when they went into his inner room, everything went dark. No audio, no video trail. Nothing. He must have meshed the entire area with copper."

Brian adds, "We haven't quite figured out how to get around screening like that, unless we use the mosquito drones."

Lori looks at them, "I need to call an emergency meeting with Admiral Whistler, General Maxwell and Commander Jackson." At the sound of his name, Kristi looks dreamily at Lori. Lori rolls her eyes in response.

Lori Sanders exits the work area of Kristi Kruger and Brian Pierce.

CHAPTER 14

TROUBLE IN PARADISE

At the Pentagon, the next day, September 25th at 04:00PM Eastern Time, Admiral Steven Whistler, General Shane Maxwell, Commander Colt Jackson and NSA superstar, Lori Sanders are conducting an emergency meeting.

Lori takes the lead, "The conclusion of our investigation is very strange. Lefty Cifer came out of nowhere, with nothing, and has built an Internet empire, overnight, through leveraging deals with the media companies. None of the terms have ever been disclosed."

Colt Jackson says, "Media companies. Why are we always coming back to the media companies? We just parachuted onto those buildings a year ago, stopping the attack on the system. How can they still be the center piece for a global winner-take-all strategy?"

The Admiral says, "I think you have your fingers on the pulse, Colt."

General Shane Maxwell adds, "Yes, he sure does. The question is, how does a devilishly, handsome young man obtain contracts, across the board, on mysterious terms, with every media company on earth?

Colt looks questioningly at the General, "Did you say, devilishly?"

"Yes…Ms. Sanders, please pull up the background we could pick up on Lefty Cifer."

Lori immediately pops up a slide on her Power Point presentation. "Yes, sir. Here is his face. Dark hair, strong features, extremely handsome and a charismatic smile, with pearly, white, perfect teeth. He radiates smoothness."

Colt raises his eyebrow in a feigned look of being hurt. Lori sees it and gets the urge to roll her eyes at Colt, but doesn't. She stifles a smile.

His first name is Lucius. Here is his California Driver's License. He lives inside of the building the General visited.

Colt queries, "Can we get ears and eyes in there, like we did at The Pearl and the Hotel Marsa Malaz?"

"Yes, we believe so. I ordered our people to put this into motion yesterday. We just need this team's approval to execute. Since the secure room is obviously protected, we will have to deploy a very similar operation there, as we did at the hotel. The mosquitos will sit and collect audio and video information and then be programmed to fly back to a retrieval site. The only problem is, we don't have a lot of time left."

The Admiral wonders, "What else can we do?" He pauses, thinking, and looks around the room. The Admiral continues, "…I'm afraid not much. Until we know more about this meeting, it really doesn't mean anything. It should make us highly suspicious, however, of anything and everything General Jacobs has told us."

General Maxwell asks Lori about the tests and simulation claims General Jacobs made. "Have you looked into his claims, Ms. Sanders?"

"Yes, sir, I have."

Colt jumps in, "And, were the tests and simulations faked?"

Lori replies, "I'm afraid not, Commander. Everything checks out. We have counter-confirmation with the CIA and they even had their very best agents on the ground when the tests were conducted. A full-scale data center didn't experience even a blip when the EMP test-blast hit them. They were completely protected. The metallic net worked. It truly is an ace up our sleeve against the Chinese, Russians and any other rogue state who might want to launch an EMP attack against the States."

Colt pays Lori a compliment, "You are always very careful, Ms. Sanders. I completely trust your judgment on this one. Something doesn't feel right, though."

Admiral Whistler puts his opinion on the table, "Until we can prove something, we have to go with our plan."

General Maxwell agrees with the Admiral. "I concur. We will have the USS Minnesota sink the three LUBEROL cargo ships out of Amsterdam.

They're controlled by QEMCO, QATAR's leading chemical company, but with the collection of Dutch and Muslim employees working in Amsterdam, the Islamic State will assume the leak came from the Dutch side. Each ship will be carrying a 2-10 kiloton warhead on North Korean SA-2 missile launchers."

Commander Colt Jackson doesn't like what he is hearing. He pushes back. "I still don't like it. There are still three tankers out of Maracaibo to account for and track. Are we seriously going to let them fire on America?"

Admiral Whistler states flatly, "Commander, as long as those ships are sailing international waters and are not attacking our sovereign borders, we cannot attack them. Is that clear?"

All eyes in the room are on Commander Colt Jackson. The Commander has just been given a warning shot over the bow, by the Admiral. There will be no cowboy stunts allowed.

Colt responds, "Yes, sir, but I really think..."

Admiral Steven Whistler, knowing how strong Colt can be, comes down hard on him, "Commander! Do you understand we have to follow orders and rely on the ability of the Metrail-mesh sky network?"

General Maxwell attempts to intercede between these two fencing officers of the United States Navy. It was clear the Admiral would win, of course, but the General tries to ameliorate the situation. "If I may, Admiral?"

The Admiral waves his hand and bows his head in a gesture to please proceed.

"Commander Jackson, with all due respect, we have national strategic security concerns about revealing our advanced defense technologies and our highly-prized assets in the Islamic State. They are poised to ransack South America, soon, and we cannot allow Croc Hunter's allegiance to us, be discovered."

The Admiral continues, "To be clear, Commander, if the other three tankers are attacked and sunk, Croc Hunter will lose his head---right on You Tube."

Colt Jackson looks at the men with a very serious gaze. It was hard to win a stare off with Colt Jackson. His cobalt-blue steel eyes seem to light on fire when he is in a tense moment. This was clearly turning into one of those. "Gentlemen, I hear you, loud and clear. You are putting the livelihood of an entire nation behind the safety of one highly-placed double agent. Let me see, that makes it about 340,000,000 people for 1. Is this what I'm really hearing, today?"

Lori Sanders turns to Colt and sits straight up, dead serious. She has a job to do and she is going to back up her boss, big time. She says flatly and talks down to Colt, "Commander, we have gone over the data and we have 100% live confirmation the Metrail mesh WORKS! Don't be thick-headed on this!"

Colt Jackson glares at Lori Sanders. He has instantly become really pissed off.

Colt holds his silent and steely-gaze and feels a growing resentment at being chided and insulted like this by Lori Sanders, right in front of his superior officers. Colt is no dummy and immediately retorts, with sarcasm:

"LEX PARSIMONIAE."

Lori Sanders doesn't understand. "Excuse me?"

Colt laughs a mocking laugh. He speaks directly at Lori Sanders with fire in his eyes. His voice rises angrily, as he speaks. "What, no computer to translate your Latin for you, today, Ms. Sanders? Lex parsimoniae---Ockham's razor. Among competing hypotheses, take the one with the fewest assumptions. Seems to me, you guys are collectively betting the future of America on the back of a system which you believe will work. That is, it will work under ideal test conditions. What if something happens? Have you ever thought of that, MS. SANDERS?"

Colt Jackson is furious at being spoken down to by Lori. He turns to the Admiral. "Sir---permission to be excused from this meeting? I have to return to base."

Admiral Steven Whistler responds seriously, "Commander Jackson, you are excused from this meeting. You will follow orders, to the letter, is that clear?"

General Shane Maxwell chips in, "I think we can wrap this up. Don't worry, Commander, we'll make sure those planes get in the air and spray every inch of the country. The conspiracy theorists will have a field day when they see the cloud formations. Little do they know, the metallic mesh is designed to protect their freedom and way of life."

Colt looks directly at the Admiral, his eyes are flashing like beacons and states with obvious disdain, "I am not in this operation. I would like to request some time off, I need to take some R&R."

The Admiral immediately grants the Commander's request, "Granted, Commander. You haven't had a good vacation since you became a SEAL. See you back after October 15. This will all be over by then."

"Aye, aye, sir." Colt looks and salutes the Admiral and General, while completely ignoring Lori Sanders. Colt storms out of the room, clearly furious. Virtual smoke is pouring out his ears.

General Maxwell turns to Admiral Whistler and to Lori Sanders. He says sarcastically, "I'd say that went well..."

The Admiral tells them, "Thank you both for your time. General, you want to stick around for a minute? Thank you, Ms. Sanders. Colt will be OK. When he gets ticked off, he steams, but it will be like water off a duck's back. A bit of time off will help him. Let's get to work."

Lori says to the Admiral, "We will, sir. Thank you both, gentlemen." She leaves the Admiral's office.

Outside Admiral Steven Whistler's Office, as soon as she is out the door, Lori Sanders runs after Colt. Colt is at the elevator, down the hallway---its door is just opening.

Lori shouts out, "COLT! COLT! WAIT! I'M SORRY! COLT!"

Commander Colt Jackson, U.S. Navy SEAL and Colonel in the United States Marine Corps, completely ignores her. He strides inside the elevator and smashes the button for the ground level. He's spits out at the elevator wall,

"I'M OUTA' HERE!"

The elevator doors close, just as Lori Sanders runs up.

Lori desperately tries to speak with Colt. "COLT!" she cries out, with the elevator doors closing and she is clearly heard by Colt Jackson. Colt continues to ignore her. Colt's and Lori's eyes meet, just as the doors close. Lori Sanders sees the fury on Colt's face and the fire in his eyes.

She says to a closed elevator door, in complete frustration, and hits the closed door.

"COLT, I'M SORRY!"

CHAPTER 15

ABOUT CARACAS

Admiral Muhammad Mubarak, Vice-Chief of Naval Staff and Admiral of the Qatari Navy, has flown to Caracas to personally meet with General Jose Lopez. The cover story for this trip is two military commanders collaborating on strategy for the protection of oil tankers, due to the threat of piracy.

At least, this is what the secretaries and the adjutants fired off to each other while organizing the meeting.

The Admiral and the General meet at the top of Mount Avila. It is filled with lush vegetation and is often called "the lungs" of Caracas. Rising above the city from the valley below, it separates Caracas from the Caribbean Sea. The park at the top of the mountain is called Waraira Repano and provides privacy for the men to speak about their plans. It is now September 26th at 02:00PM in Caracas, 02:30PM Eastern Time.

General Jose Lopez is walking with the Admiral, speaking to him, "Venezuela stands ready. The oil tankers contain the missile launchers. We have also copied the packaging design so graciously given to us by you. This shipment will be encased in lead and concrete, with a six-foot swimming pool wrapped around it--on all sides."

Admiral Mubarak replies, "You have done well, General. Caliph Abu-Bakr and the Emir are very pleased. I guess the only question is, are there any changes? When and where?""

The General answers, "We will fire them from the Gulf, outside of New Orleans, for maximum impact across the oil, gas and defense infrastructure of the deep south. Another ship will be sent to Charleston, where everything from Florida to Washington D.C. should be devastated by the EMP blast. The last tanker will head to the Long Beach Terminal in L.A. Sent in different directions, the EMP affect will be devastating to the natural gas, oil, defense and entire infrastructure across the west, South and Southeastern portion of the United States. "

The Admiral is very pleased, "Outstanding! When will they depart?"

"The longest trip is to Long Beach. It is over 3,000 miles and will take nearly eleven days. Ten days and twenty-one hours, to be exact. It is going to depart on September 28th---giving it a couple of extra days to get in place. Next, the tankers for New Orleans and Charleston will set sail. They will depart together, at dawn, on October 4th---giving them plenty of time to arrive at their destinations. Their journeys are only four to five days long. They will both sail together, until they reach the Antilles Ridge, right between Cuba and Haiti. From there, one will make its way north to Charleston and the other to the Northwest, to New Orleans. Everything is set to launch on the 11th, on Yom Kippur, as planned."

Admiral Mubarak smiles, "This is very good news."

General Jose Lopez asks, "What about LUBEROL, are those vessels set?"

"QEMCO has made the necessary arrangements and has ordered for their shipment. The Dutch brothers have placed the three missiles on three different vessels. Their journey is nearly 10 days. They all leave Amsterdam on September 30th."

Both men look at each other and smile. They give each other a customary hug and pat on the back.

The General states, "Once this mission is complete, I look forward to meeting again in your fine country. Your service to Islam and the true faith is a gift from Allah, himself."

Admiral Mubarak replies, "We are only doing Allah's will, as are you. By the way, the Emir wanted to know how his gift is working out."

"Please tell his majesty I am forever grateful and in his debt for such a beautiful set of gifts. Each and every one of them is precious to me."

"Excellent! And the special gift? Fatima? The Emir knew you fell for her at the Marsa Malaz and he wanted you to know he remembers those who are faithful to him and our cause."

"Admiral...PLEASE tell our dear Emir my heart leapt out of my chest, like a young deer vaulting over a fence, when I saw her. She is exquisite and services me, best of all."

Admiral Mubarak frowns, "The others are not as good? The Emir will be most upset!"

General Lopez holds his hands up, waving them emphatically, "NO, NO, NO!! They are all PERFECT in every way. Like diamonds, they all have a slightly different sparkle."

The Admiral laughs in understanding. "The Emir will LOVE to hear this. It will give him joy to know he has made you happy. Your bravery is now a growing legend among all of us. First penetrating the highest level of a Catholic government and being the point of our spear in the launch of ISSA---the Islamic State of South America!"

The General responds, "I am only a humble follower of our great prophet. I only choose to do the will of Allah. Allahu Akbar!"

"Allahu Akbar! Good luck, my friend. I will return now to Qatar and tell the others, in person, everything is set. We cannot risk electronic communications, yet, until we knock out the Americans. The NSA is everywhere!"

General Jose Lopez speaks to Admiral in Arabic, "As-salamu alaykum." (May the peace and mercy of Allah be with you).

The Admiral returns the greeting, as is their custom. "As-salamu alaykum. Allah ysallimak." (May the peace and mercy of Allah be with you. May Allah protect you).

General Lopez responds in kind, "Allah ysallimak."

The men depart the park and go down the mountain via the "teleferico," the gondola system which whisks visitors down to the city of Caracas, two kilometers below. They each take separate teleferico cars, with the Admiral departing first. The ride takes only 15-20 minutes and covers 3.5 kilometers.

Caracas, Venezuela

Las Mercedes, Caracas, Venezuela.

Photo Credit: Guillermo Ramos Flamerich, his own work.
Date: 28 June 2008, Source and license:
http://commons.wikimedia.org/wiki/File:Las_Mercedes_-_Caracas.jpg
http://commons.wikimedia.org/wiki/Commons:GNU_Free_Documentation_License, version 1.2

CHAPTER 16

AIR AMERICA

On the same day, September 26th, 3,180 miles away in Tucson, Arizona there are many boneyards. Most of them remain undisturbed from Old-West days, where heroes and villains were gunned down and buried, some right in the middle of town.

However another "boneyard" just outside of Tucson, is where civilian airplanes go to rest when they are taken out of service. It is called Marana Air Park, which was formerly "Air America's" operating base for the CIA during the Vietnam War. Hundreds of aircraft are left standing there. The dry desert heat ensures they don't rust and this provides a weigh-station for older aircraft to be marketed, sold, scrapped and refurbished.

Marana, is also called Pinal Airpark, and has a highly secure airfield. It is home to a host of ghostly-white, very large, modern, wide-bodied aircraft. Each of them is capable of carrying tens of thousands of gallons of Chemtrail mixtures into the sky, without raising public suspicion from this location.

General Charles Jacobs is personally reviewing the aircraft for the operation and is meeting with two Chinese-American employees of the company which maintains the aircraft, under an outsourcing agreement.

General Jacobs greets the men, "Derek, Charles, good to meet you." He extends his hand in a firm handshake.

Derek Wong replies, "General, the Master has spoken so highly of you. We are honored to meet you." He bows his head in reverence.

His co-worker, also a devoted disciple to Lefty Cifer, Charles Chen, also bows and says, "It is likewise a pleasure, sir."

"Our allegiance is to our Master, the mighty Lucifer. We successfully got ourselves assigned here and we have used our cell phones the last few weeks to call various government agency numbers in China, engaging in conversation with the receptionists for a number of minutes before hanging up."

Derek Wong explains to the General, "The trail will lead back to us, after the damage is done, and the trail will lead back to China. This creates another potential conflict for the American government. The more chaos we can create, the better it is for our Master."

The General is satisfied and happy, "Excellent job, men. Just Excellent. Your associates, with a little help from me, are ready at Altus Air Force Base, where the other Chemtrail team will deploy. They will head south and east---your aircraft will cover the north and west. The pilots know nothing. They believe they are just on an aggressive spraying mission. It is vital you switch the Metrail mixture with the non-metallic aerosol mixture, prior to getting in the air. By the morning of October 11th, the entire sky of the United States will be filled with strange, white, puffy clouds which will do absolutely nothing to stop the EMP attack."

"The plan is masterful General," says Derek. Charles bows his head, again to the General, in reverence.

General tips his cap to them, "Gentlemen, good luck. When the planes take off, finish your shifts and act like everything is normal. Drive north and head to the Master's lair on Park Boulevard in Palo Alto. If you drive all night and into the morning, you will be exhausted, but you will avoid the effect of the EMP. The aircraft slated to fly over the Bay Area and Northern California need to be filled with the 'real thing.' Make sure it is loaded with the real Metrail mixture. We need to protect our Master, Lefty Cifer."

"Consider it done, sir! We will see you, again General, at the lair. Thank you for this honor and great mission, sir." Derek responds for both men.

"You are welcome. Goodbye gentlemen." The General departs.

CHAPTER 17

THE PAULO PLAN

Meanwhile at the Naval Amphibious Base, Little Creek, Virginia, Colt Jackson is on vacation. He is at the Navy SEAL base, located at Virginia Beach. It is now 8PM on September 26th on the East Coast.

The warm September evening allows Colt to walk casually along the seashore, in his shorts. The waves wash over his feet.

He is on a vacation, after all…

Colt has been doing a lot of thinking the last couple of days since the intense meeting with the Admiral, General and Lori Sanders. He dials a number via an encrypted connection. He continues his stroll along the beach, alone in the evening's warm breeze.

Colt speaks into his cell phone when the other party answers, "Hello, Croc Hunter?"

Inside the home of General Jose Lopez, it is 7:30 PM Caracas time. The General recognizes the voice and says instantly, "Is that you, Papa Bear? Good to hear your voice, old friend. Catch any mosquitos lately?"

Both Colt Jackson and the General share a laugh. The General immediately walks out the sliding glass door to the terrace, outside. He closes the door behind him.

Fatima watches him carefully. The conversation continues between the two men in private.

"Croc, I need to speak with you off the record. I need your help. But, before we speak, I need to speak spy-to-spy. I need your word this conversation goes to no one else."

"Papa Bear, you know you have my word of honor---on Allah, himself."

Colt replies, "That's good enough for me, Croc. Here's the situation. I know you are launching the tankers, three of them as planned. I was in a top-level meeting with players---not to be named. We uncovered some very disturbing information about one of the parties tasked with defending the U.S. from the attack. Without going into details, suffice it to say I have little, or no confidence, in what is happening. The entire country, in my opinion, could be open to a complete Pearl Harbor-like attack."

The General is shocked, "How is this possible? I thought certain technologies were being deployed to defeat the threat."

"I believe the technologies, properly deployed, would work, Croc. The problem is, the assumptions being made assume a perfect execution---just like in test conditions. It leaves no margin for error for technical failure, human error, or God forbid, espionage."

"ESPIONAGE?"

"That's my real fear, Croc---enemies from within. We picked up some G-2 which is really, really fishy. We can't prove anything and, you should know, I have been personally ordered by both Admiral Whistler and General Maxwell to stand down, 100%. In fact, officially, I am supposed to be on vacation until October 15th, to let this all blow over. My special team isn't even to be anywhere near any part of this operation."

"How is this possible? This definitely doesn't make any sense. Why??"

"Croc, they're trying to protect you. They don't want a direct attack on your three vessels, because you are, without doubt, one of the most important foreign intelligence agents working for the country. You are vital to controlling the spread of radical Islam."

"I see. And you think this is important enough for me to blow my cover, is that it? If it means saving the country from such a potentially devastating attack, I am more than willing to blow my cover!! An attack like this would send America back to the Stone Age---I kid you not!"

"Thank God you understand. What would you say if I told you I have a plan to eliminate most of the threat, with resources I can deploy on my own---*and* keep your identity and mission secret?"

The General says, "I would say, I am all ears, my friend."

Colt explains his plan to the General.

Back at Colt Jackson's Georgetown Park condominium at 1080 Wisconsin Avenue, it is now a day later, September 27th at 6:00 PM. Colt has returned from his visit to Virginia Beach. He left his regular cell phone in his condominium the last three days, while he made his excursion to the SEAL base at Virginia Beach.

He looks at the phone and sees he has 27 missed calls---20 of them are from Lori Sanders. He totally ignores them.

He is on vacation, after all…

Colt likes to think of this as a "working vacation" and has called a special meeting for seven of his most trusted men.

Present in his condominium are Lieutenant Cody Custer (call sign "Tiger Fox"), Chief Ryan Duffy (call sign "White Shark"), Chief Spider Sinclair (call sign, "Black Widow"), Master Chief Bull Sherman (call sign, "Sierra Hawk"), Ensign Dizzy Lee (call sign, "Stone Fish"), Master chief Jack Flintlock (call sign, Killer Whale") and Petty Officer Duke Moran ("call sign "Stingray Charlie").

They all sit in his living room. Everyone was instructed to leave their cell phones at home. No one was going to track the whereabouts of these men on this evening. With everyone collected in his condo, Colt gives them a hand signal. No one speaks in Colt's home. All of them are heading to Paulo's. After Colt's phone call to its owner, he agreed, as a personal favor to Colt, to close the restaurant this Tuesday evening between 6:15 and 7:15PM.

Paulo's restaurant is at the corner of Wisconsin Avenue and N Streets, in Georgetown, and is only a few blocks from Colt's home.

The men walk together, up Wisconsin Avenue and approach Paulo's Restaurant on the east side of the street. Paulo's quaint "A" frame exterior greets them and the men walk inside.

The restaurant's warm and inviting interior, along with its delicious Italian food, is a mainstay of Colt's and pretty much all of Georgetown.

The men head to the back of the restaurant and everyone takes a seat.

Colt asks the owner to make sure the staff are given an hour off. This group would not be ordering food for this meeting.

The owner, an old friend of Colt's, gladly agrees. He was a veteran and would do anything for a Medal of Honor recipient. He also understands this meeting is important for Colt to keep secret---and off the books.

Vittorio Luigi Lombardi, owner of Paulo's speaks, "You and your men make yourself at home, Colt. I am going to take off and run an errand, just up the street. I'll be back by 7:15PM. If you get hungry, you know where the kitchen is!" He winks at Colt.

Colt smiles and thanks, his buddy, "Ahhh, Vittorio, how can I ever repay you, my friend?"

Vittorio Luigi Lombardi responds warmly, "Your friendship is enough. If you guys leave early, just lock the door behind you. I have left a sign on the front door telling patrons we had a problem in the kitchen with the electricity, but it will be turned back on by PEPCO at 7:15PM."

Colt cracks a half-hearted attempt at a joke, "That's a pretty confident 'turn on' time for the Potomac Electric Power Company, Vittorio!" He winks and laughs.

Vittorio Luigi Lombardi starts to leave, but turns around, "Oh, I forgot to tell you I wrote at the bottom of the note---'I personally threatened PEPCO with death, if they didn't finish on time!'"

He laughs and all the men laugh, together. Vittorio shakes Colt's hand and looks at the gathering of faces. "I don't know what's on your mind, gentlemen, but I wish you the very best of luck!"

All the men stand up out of respect for Vittorio and salute him.

"HOOYAH!---VITTORIO LUIGI LOMBARDI!"

Vittorio departs and locks the front door of the restaurant.

Colt Jackson turns, now dead serious to his men, "OK, men. Now we can finally talk. Here is the deal. Right now, I have to ask each of you two things. The first will be easier to answer than the second. First, I need to speak to you in complete confidence. NO ONE AND I MEAN NO

ONE, outside of this room can know this meeting took place, or you were at my condo tonight, and they can't know anything we discuss this evening. If you can't do this, there is no shame. NOTHING will change between us. The back door of the restaurant doesn't require a key to unlock and you can make your exit there. I will turn my back to make it easier for you, if you cannot agree to the first request for complete secrecy."

Colt turns his back and walks toward the front door, giving his men some time to confer with each other. Colt waits several minutes and then returns to the back of the restaurant. All of his men are still seated.

Lieutenant Cody Custer starts it off for the SEALs, "What took you so long, sir? We all knew our answer right away. We're on board 100%."

Colt nods his head in thanks, but then continues, "Not so fast, Tiger Fox. I still need to ask you the second question. I cannot go into details until I know you are in, but I am asking if you men will accompany me on a mission---which is totally unsanctioned by our superiors. Furthermore, the mission is highly dangerous, will put us in the air and at sea and we will be consorting with a country many people today consider a clear enemy of the United States. If knowledge of our involvement gets out, we will be imprisoned for disobeying a direct order and potentially for conspiring with the enemy---and probably a hundred other charges the military prosecutors will dream up. They are good at that. I am going to turn my back one more time. This time I'll leave for five minutes---and go take a leak. Oh yea, one last thing. If the mission is successful, we might just save the country. Take your time, gentlemen."

Cody Custer looks at the group of SEALs around the table, "Well, gentlemen. What do you think? Shall we go around the table? I think everyone deserves to speak on this one. Master Chief? Do you want to start?"

Master Chief Jack Flintlock says, "I'd follow Papa Bear to Hell and back. I'm in."

Chief Spider Sinclair blurts out, "This is a no-brainer. Colt Jackson has pulled my ass out of the fire so many times, I've lost count. I'm in."

Chief Ryan Duffy affirms his brothers' sentiment, "Papa Bear has been our brother, our chief. All for one and one for all, right? I'm in."

Master Chief Bull Sherman says, "You guys don't know this, but my family's name is the one the Sherman Tank is named after. My great-great-great Grandpa was General Sherman from the Civil War. Loyalty is everything in our family. I'm in."

Ensign Dizzy Lee is speaking in his distinctly Southern Drawl, "Well, I have to keep my family name up, too. Robert E. Lee backed Stonewall Jackson and this Lee is sure as shit going to back Commander Colt Jackson, even if he is a damn Yankee!" He smiles at the team.

Petty Officer Duke Moran looks at the group seriously, "I'll never forget Papa Bear going back for Ocean Dragon and facing that Tiger shark, mano-a-mano. He would do anything for us, when the chips are down. I'll do anything for him, also. Stingray Charlie is in."

Lieutenant Cody Custer looks around the room and catches every man's eye, "That leaves me, Tiger Fox. What can I tell you guys? He's kicked my butt in SEAL training, he's been my mentor, he's rescued me from Priest Assassins and he's given me command. I'd be a complete ingrate to not back this man one million percent. It's unanimous, then. Good, I'll go get him. No point in wasting time."

Colt is facing the urinal, just finishing his pea. Lieutenant Cody Custer opens the door and looks at his Commander, pissing into the ceramic latrine.

"Sir? We don't need five minutes. We're all in."

Colt responds to Cody, "Well, I need another 30 seconds---oh yea…Cody?---Thanks."

Lieutenant Cody Custer responds to his commanding officer and friend, "You're welcome, sir. It was a no-brainer for each and every one of us. Killer Whale said he would follow you into Hell and back, sir. We all feel this way."

Colt Jackson is trying to hold back tears from this honor his men have given him. He doesn't turn his head from the wall.

"I'll be right in. I promise---you won't be bored."

A couple minutes later, Colt returns to the table. "Men, thank you. This is the situation. First of all, every one of you is going to be on "vacation" from October 4th through October 7th. This is when our op goes down."

Cody Custer says, "What continent, sir?"

Colt replies, "We're going to be protecting this one. There is an imminent, confirmed attack on the U.S. with nuclear weapons." Colt pauses for effect. The men sit, stunned

.

"The attack is being directed by the Islamic State and based upon our surveillance, this top secret information was given to the President, the Admiral, the NSA and CIA. Specifically, ISIL has created and planned an elaborate EMP attack on every coast of our country."

Chief Ryan Duffy, "Which date is the launch, sir? Do we know the exact day?"

"Well, none of those days are the launch date. The attack is planned for the 11th of October on Yom Kippur. It will come from three oil tankers out of Venezuela. They will be in route during our "vacation" to their ultimate launch-site destinations. The ships have been modified to hide and mask the weapon's nuclear signature in their cargo bays, making it impossible for our satellites to pick up any radiation trail."

Ensign Dizzy Lee speaks up, "Sir, why not bring in the Admiral? He's tight with you, right?"

"I can't do it. Due to national security reasons, he specifically and personally ordered me to stand down during this time period. I nearly let my temper get away from me in a meeting I had with him and the NSA. I was furious they wouldn't order our Team to simply eliminate the threat by sinking, or otherwise attacking the vessels---long before they are sitting off the coasts of the USA."

Petty Officer Duke Moran doesn't understand, "It doesn't make any sense, sir. What piece are we missing?"

"Men, they are relying on a proven system to repel EMP attack. It is top secret, but they are willing to use this system and alert the Russians and Chinese to its existence, rather than compromise a different and important national security secret. The only thing I can tell you is it's a highly prized foreign asset, critical to bludgeoning the Islamic State's interests."

Master Chief Bull Sherman says, "Papa Bear, what do you want us to do?"

"Chief, I've already been able to speak directly with some of our buddies in the stealth chopper pool. I asked them the same questions I asked you. When they learned what was up, they are going to give us the best of the best---the new Sikorsky S-97 Stealth Raider. They are implementing and creating a drill where they will be able to fly off into the ether for 48 hours. They have named the drill, 'Deep Ocean Danger.'"

Cody Custer asks, "Any significance to that name, sir?"

Colt answers, "Well, to insert us in our GPS drop zone, the choppers need to exceed the limit of their flight range and will be flying over ocean which is 10,000 feet deep. I'm pretty sure this is how they came up with the name."

Eyebrows go up among some of the men.

Ensign Dizzy Lee, the tech-savvy, smart SEAL, who was Ocean Dragon's friend says, "Sir, I read about the Raider. It is supposed to have a range of greater than 600 Kilometers and its X2, double-bladed technology, broke the world speed record for a rotary craft. It went over 260 knots--nearly 300 miles per hour, from what I read. Just how far are we going?"

Colt replies, "Just one way of the journey is over four times that distance. We're going 1,600 miles to our targets' GPS intercept point."

The Ensign scratches his head, "1,600 miles? I don't think the Raider can put on exterior tanks, or an inner fuel bladder, to make it that far. Its range is only 370 miles, without an auxiliary tank---is it over ocean?"

Colt replies, "A significant part of it is, Ensign. Our flight path will hop down, along the U.S. East Coast to the Naval Air Station in Ft. Lauderdale. That will be our last over-land refueling stop."

Ensign Dizzy Lee quickly does the math in his head, "Sir? Sorry to be a pain, but doesn't that still leave 650 miles to cover---over water?"

Master Chief Bull Sherman can't resist the urge to be a wise-ass, "I guess we better make sure we bring along our Shark Shields, huh, sir?"

Everyone laughs.

Colt looks around the table, "All kidding aside, yes, Master Chief, bring those things. They really work and we will certainly need them where we're going. The Stealth Raiders are top secret and we can't be seen on this mission. Therefore, we will not be able to fast rope down onto the deck of the tankers. Furthermore, we have to make drops for two separate teams, but at the same location. Half this team will take out one tanker and the other half will take out another. The tankers will be traveling together."

Colt turns to Ensign Dizzy Lee, also known to his commanding officer as "Stone Fish."

"To your point, Stone Fish, we have been able to obtain the cooperation of the Venezuelan Navy to have its frigate, the Yakuana, leave its safe-haven port in Havana and provide refueling for the Raider off the Bahamas."

At the mention of Venezuela, all the men's eyebrows rise, again.

Cody Custer speaks up, "Sir, I learned a long time ago to not question you, or your judgment. If Venezuela is our ticket in and the crews' ticket home, then so be it."

Commander Jackson looks at his Lieutenant and says, "Thanks, Lieutenant. Gentlemen, this is something you are going to have to trust me on. The Raider and you men will never officially be on the Yakuana, but you have my word of honor, the landing will be protected and the ship's crew will be on lockdown. Only the Captain of the ship and a specially trained agent will have knowledge of the landing. The agent will assist us in refueling the Raider---both coming and going. The Captain will be running a mock contamination drill to clear the decks. That is all I can say, at the moment."

Cody looks at his Commander and says, "How'd you pull that off, Papa Bear?"

Colt smiles at Tiger Fox. "Let's just say I pulled in a few favors from a very good man."

All the men together cry out, "HOOYAH, SIR!"

Colt responds, "Excellent! OK, make sure you bring the Sealskins from our last mission at The Pearl. Those things worked like magic and we'll need them to get on the tankers undetected. We'll use two of the new stealthy CRRC's*---the new Zodiac boats---which we'll also deploy from the Raider. The Raider's six seats are being ripped out---we'll sit on mats and will rest on top of the CRICS. Their footprint is small--they're each only 2'6" x 4'11", when stowed, but it won't be a comfortable ride, I'm afraid."

Chief Spider Sinclair takes his turn with his sarcastic wit, "Sounds like we get a six-to-seven hour nap during our taxi ride, sir---it'll be just like the Ritz." Everyone laughs around the room. "…But, sir, once we board the tankers, what do we do?"

"That's a great question, Chief. First of all, unless we are being engaged in mortal combat, do not engage. If you have to take someone out, we MUST take them with us and throw them overboard---without being seen. Remember, you'll be invisible, but the dead person won't be."

Lieutenant Custer asks, "What about the nukes?"

"They will be sitting on top of SA-2 missiles from North Korea. Each ship has exactly one nuke on board. Our mission is to booby trap the launch system. When they light the fuse, we want the missiles' motors to become destroyed, by a minor explosion. We will use a shaped charge to make sure we don't blow up the liquid fuel in the upper stage, or the solid-fuel booster, in the lower stage of the missile. The blast should be big enough to set off heat and sound signature warnings for U.S. satellites, as well as the sonar detection systems deployed throughout the Caribbean---but not enough to sink the tankers. That racket will cause the U.S. Navy and Coast Guard to come down on those ships in a heartbeat."

Master Chief Jack Flintlock connects the dots, "So, we don't want to leave any fingerprints behind, huh, sir?"

"You got it, Chief. We will vanish like ghosts, meet the Raider at our GPS pickup point, and the last man to be picked up, which will be me, will scuttle the Zodiacs."

The Master Chief answers, "Gotcha, sir."

Colt continues, "Gentlemen, throughout this mission, you will keep your Sealskins' active camouflage capability switched on. The powers that be

* Nicknamed, "CRICS"

will be pouring over satellite feeds to see what happened to those tankers and we cannot have any of us being identified. They may see the Zodiacs docked on the tankers, but they won't see any of us. What they can't see, they can't prove. Are we clear on this, men? Nevertheless, except on my special order, when we board the tankers, you are to keep your faces down. We don't want to give them any help with their facial recognition software. Are we clear?" Colt looks around at all the men.

They immediately answer, "CRYSTAL, SIR!

Lieutenant Custer is thinking out loud, "The ArmorFlate bullet proof material on the Zodiacs will protect us, if the tankers happen to spot us and start firing. I'd love to be a fly on the wall if a crewman looked down and saw an empty Zodiac snuggle up next to the ship. They'll be pissing their pants! They'll think pirates of the Caribbean ghosts are steering toward them."

The room fills with laughter again.

Colt Jackson likes this thought. "Hmmmm, Pirates of the Caribbean---That's a great idea, Tiger Fox."

Cody Custer asks, "What do you mean, sir?"

"I'll let you know in due course. Don't you worry though. I promise we'll have a little bit of fun on this op!"

Colt whips out a map of the Caribbean. It is a close up of the Eastern tip of Cuba and the west coast of Haiti. He points to an "X" on the map. "This is our rendezvous point with the tankers. From Maracaibo to either New Orleans, or Charleston, they make their turn starting about here---right between Cuba and Haiti. The water off of Guantanamo, along the Greater Antilles Ridge is very deep. At our drop point, it is exactly 10,113 feet deep. This is the Latitude and Longitude:"

	Latitude	Longitude
Rendezvous Point	19° 48' 11.22"N	74° 04' 01.11"W

Ensign Dizzy Lee knows exactly where they're going, "That'll be only about 50 miles from Gitmo, sir. Too bad we can't just make it back there and hitch a ride home out of Guantanamo Bay."

Colt answers, "Under different circumstances, we sure would Ensign. But, we'll have to get back on the Zodiacs and head to the pickup point. We'll signal the Raider when we're on our way. It'll be there waiting for us. Any more questions? You are now all going to be on an official vacation from October 4-7…I hope you enjoy it," Colt says with a light furrow of his eyebrows and then smiles at his men.

The men automatically reply, "HOOYAH, SIR!"

Chief Petty Officer Jack Flintlock, with his naturally quick wit says, "Who can beat a trip to the Caribbean for a little R&R, right sir?" He smiles a huge grin and the room cracks up in laughter.

Lieutenant Cody Custer lifts his hand, "Sir, I do have one question."

"What's that, Lieutenant?"

Lieutenant Cody Custer says what is on everyone's mind. "What about the other nuke? What about the other tanker?"

The room goes completely silent.

Colt answers, "Tiger Fox, I'm sorry to say, I have run out of bullets on that one. This is the only team I could truly trust and count on to not spill the beans. If the beans get spilled, we'll all be in the brig and these nukes will be purposely allowed to go off over the homeland. It is an unacceptable risk. I'm sorry to say I have done my very best to avoid a potential calamity, but it is not enough, I fear."

Lieutenant Custer asks, "Where is the third tanker heading, sir?"

Colt Jackson replies, "It's heading through the Panama Canal on a direct trip to Long Beach Terminal---Los Angeles."

The room is deadly serious and every face is fraught with concern.

Colt continues, "I hope the defense layer they are planning on using works---that's all I can say. Gentlemen, whether you believe in God, or not, and you know I believe Jesus Christ is my savior---pray for those people over L.A. They may surely need it when October 12th comes around, in the aftermath of the blast the day after. It will be chaos."

Master Chief Jack Flintlock speaks out loudly, "Holy shit, sir. There's nothing we can do??"

Colt answers, "I have done everything I can, Chief. Everything---even to the point where I will probably be summarily court-martialed, when this is over."

A complete silence sweeps over the room.

"Gentlemen, thank you. Thanks to each and every one of you for being true. You are a rare breed. I'm so proud of you. It's time to take off. I'll see you in a few days, gentlemen." Colt Jackson salutes and all the men exit Paulo's.

Colt locks the door behind him and leaves the key on top of the right front tire of Vittorio Luigi Lombardi's fire-engine red, Cadillac convertible.

Georgetown, ever short of parking spaces, never causes Vittorio a problem for his prized vehicle. It sits prominently right in front of Paulo's, with six inches of his car's bumper hanging over the crosswalk. In all his years owning Paulo's, Vittorio has never gotten a ticket. The cops know where to go to get a good meal.

Colt speaks up, "This is Vittorio's car. He's quite a character. You should hear him with the top down singing to Dean Martin "That's Amore" and then he switches over to Leonard Cohen. The chicks in Georgetown always give him a close look. He loves it."

Everyone looks at the car and laughs.

Chief Ryan Duffy looks down at the front of the bumper on the car, "Did you guys see his license plate? It actually says, "MI AMORE," He seems like a pretty cool guy, Commander."

Colt Jackson replies, "He is, Chief. He is. When this is all over, we'll have to come back and have a party---on me."

"HOOYAH, SIR!"

"Good night, gentlemen." Colt waives good bye and walks down Wisconsin Avenue towards his Georgetown condominium.

The men watch from the light pole, right outside Paulo's, as their Commander walks down the street and turns down the bend toward M Street. No one leaves the sidewalk in front of Paulo's until they see their leader disappear---lost through the trees and crowded sidewalks of Georgetown in the early evening.

Lieutenant Cody Custer looks at the men around him, "That, gentlemen, is one hell of a guy."

"HOOYAH, SIR!"

The men then go their separate ways for the evening.

CHAPTER 18

RAIDING ON THE RAIDER

On October 5th it is 07:00PM Eastern Time. Sunset was at 06:40PM in Virginia Beach and Commander Colt Jackson assembles his men at the helipad, under the cover of nightfall. The Raider is loaded and ready. The men found some ultra-thick foam rubber and made a flat area to rest and sleep on---placing the matting on top of the Zodiac containers. Their upcoming seven-hour journey will bring them over water 10,000 feet deep, before they will launch them out of the chopper and into the sea.

Lieutenant Cody Custer turns to Colt Jackson and shouts over the noise of the rotating blades. Even a stealthy Raider makes noise when it is on the ground. "Do we have confirmation the tankers left, as scheduled?"

Colt shouts back, "Affirmative. Both ships departed yesterday at dawn. It will take them a total of forty-five hours to reach our GPS point off Guantanamo Bay."

"Perfect, sir. We'll be able to board them in the middle of the night. Even with the Sealskins lit up and active, it is nice to go in under the cover of darkness."

Master Chief Jack Flintlock pipes in, "I guess that was all just luck, huh, sir?"

"I'd rather be lucky than smart, any day, Chief, but, like Henry Ford, I've found the harder I work, the luckier I get!" To the seven SEALs, Colt orders, "Saddle up, men. Let's go!"

They all climb aboard the Raider and it takes off.

Inside the Sikorsky S-97 Raider Stealth Helicopter, the men are surprised as the helicopter rises swiftly in the air and the pilot switches it to "whisper mode." The men look at each other and realize this is the first time they can actually talk to each other, in a helicopter, and not have to shout.

Ensign Dizzy Lee says, "Unreal. This thing is super quiet." He looks around the interior of the brand new helicopter. "Wow. Way cool. Sir?---Permission to go up to the cockpit and look around?"

Colt Jackson smiles at his techie and says, "Granted, but make sure the pilot is OK with it---and don't touch anything, Stone Fish. I know how you like to tinker with new toys!" Colt winks at him.

"Sir, yes sir!"

The trip goes by quickly. It is now 01:00AM on October 6th. The Raider slices through the air with ease. The men rest and aren't even bothered when the chopper stops to refuel along the coast. They are now past the Bahamas and are at the rendezvous point to meet up with the Venezuelan Navy Frigate Yakuana.

The chopper is 285 miles southeast of Miami. There, they refueled at the Naval Air Station in Ft. Lauderdale. The men are now wide awake and can see the ship's lights, below. The Captain of the Yakuana flashes the ship's deck lights. As the Raider approaches, a lone sailor is signaling the Raider with his flashlights to land on the ship's helipad. They've covered about 1,400 miles in about six hours, including refueling stops. Their air speed has been a constant 243 knots per hour, or about 280 mph. This bird can really fly.

Colt Jackson sticks his head into the cockpit to speak with Naval Aviator Pilot, Commander James Forrestal. "Hey, James, I just want to say thanks. I know you could lose your career over this one. I definitely owe you a beer."

Commander Forrestal replies, "Colt, when Prince Harry called and told me what you needed, there wasn't even any question. I served in Afghanistan with that brave son-of-a-bitch, Harry. I was being chased by some Pakistani jets and was dead in their sites, but Harry came up outta' nowhere and took them out with his Apache. We became buddies after that. He also told me you saved his ass over there. So, as our buddy the Prince would say, "No worries, mate!"

Colt slaps him on the shoulder in thanks. "Thanks, James. I'll never forget it and your country owes you a debt of gratitude."

The Commander responds, "That's good enough for me, Colt. Cheers.

I'm putting her down on deck now. We'll refuel and are less than an hour from our GPS drop zone for the tankers. I'll confirm the location of the vessels via satellite and radar. Then, I'll put you down only about a mile from the ships. They won't see us and they certainly won't hear us. You can take the Zodiacs from there to both ships."

"Roger that James. I'm going to brief my men, but Commander---you're the man. We couldn't do this without you. We'll contact you on the encrypted channel when the mission is complete and take the Zodiacs to the pickup point. You'll be safe to go back to the Yakuana, which will continue to steam toward us."

Commander Forrestal says, "I made sure the auxiliary tanks were full, just in case. Again---no worries mate! Good luck."

At 01:10AM, Commander James Forrestal smoothly sets down the Stealth Raider on the deck of the Venezuelan Naval Frigate Yakuana. The only person visible on her deck is the agent, inserted by General Lopez. The men onboard the Raider can see him hurrying to bring the hose to the chopper and refuel it. The SEAL team scrambles out to help.

They refuel in less than five minutes and the Raider takes off again, into the darkness.

By 02:15AM, at Latitude 19° 48' 11.22"N and a Longitude of 74° 04' 01.11"W, the SEALs are over 10,000 feet of water which covers the Greater Antilles Ridge. This is their drop zone, right between Cuba and Haiti in the Caribbean Sea.

On Colt's mark, the SEALs' drop their Zodiacs' into the water. The men jump out the door and the Stealth Raider immediately takes off.

The men hit the water and swim to the rubber rafts. Out of their "footprint," the boats magically spring to life and inflate in just 2 minutes. The men climb in. The SEALs can see they are precisely one mile from the tankers and can plainly see the ships' lights and outline on the water.

Colt Jackson takes command of one group and Cody Custer takes command of the other.

Colt says, "Tiger Fox---good luck. Good luck men. Don't get caught!"

Lieutenant Custer replies, "Aye, aye, sir. We'll stay in touch over the COMM line."

"Roger that." Colt signals everyone to activate their Sealskins. The men disappear to anyone who might look at them. Their robo-vision, Q-Warrior headgear, of course, allows them to see each other through its thermal imaging capabilities and the special ultra-violet tagging on their suits, via its integrated, multi-purpose lenses.

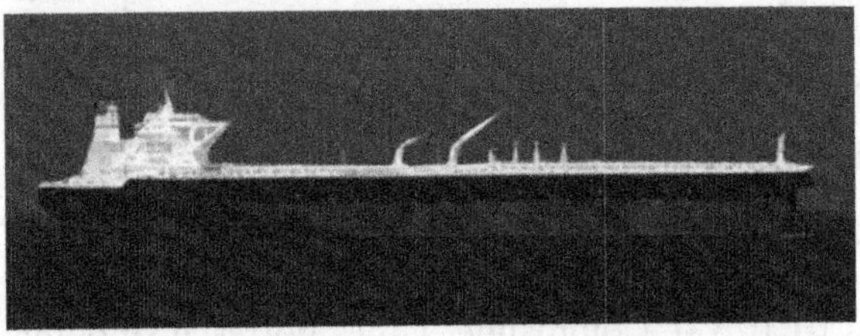

The Commercial Oil Tanker "El Toro"
Source: Wikipedia. Public domain image of oil tanker AbQaiq
http://en.wikipedia.org/wiki/Oil_tanker#mediaviewer/File:Supertanker_AbQaiq.jpg

At 02:20AM Eastern Time, both Zodiacs speed toward their targets. The enormous tankers approach, traveling at 15 knots, while the speedy Zodiacs, with their twin 55 horsepower engines, do 18 knots, or 21 miles per hour, over the open water. With the ships and the Zodiacs approaching each other, they close their range at what seems like twice the speed. The SEAL teams each make it to their respective ship targets in about two minutes.

Over the COMM to Papa Bear, Ensign Dizzy Lee says, "Sir, these radar absorbing materials work great---no signs of alarm on the tankers from our approach."

"Copy that." Colt clicks the COMM to Tiger Fox, "All clear, Tiger Fox? Any alert?"

Lieutenant Cody Custer replies, "No sir, none. We're just snuggling up next to her now. We'll be on board in minutes."

"Roger that. We're on the same timeline. Good luck."

By 02:24AM, the respective SEAL teams each shoot a harpoon rope over the top of the tankers' deck walls. They pull them tight, tie them to the Zodiacs and climb quickly up the rope. No one is on deck in the middle of the night. Each team makes its way to the precise location of the missile launchers, another gift from Croc Hunter's excellent intelligence.

On board the Venezuelan oil tanker "El Toro," Commander Colt Jackson, Ensign Dizzy Lee, Master Chief Jack Flintlock and Petty Officer Duke Moran arrive at the bay where the tanker has been modified to hold cargo. The all look at the large, North Korean missile. It is poised, ready to launch.

Ensign Dizzy Lee says, "Papa Bear, I've never doubted you, but if I ever do, you have permission to kick my butt. How did you know exactly where to go for the missiles?"

Colt answers, "Ask me no questions, I'll tell you no lies, Stone Fish. Let's get to work."

The men look again at the North Korean SA-2 missile and shudder. A crane has pulled the concrete cocoon off of the missile. The cocoon is just hanging above the launch pad, directly over the rockets and its deadly nuclear payload.

Master Chief Jack Flintlock comments, "Clever bastards. Look sir, they're shielding the signature of the nuke from our satellites---the nukes will be invisible all the way until this ship reaches New Orleans. The concrete cocoon is shielding it!"

Papa Bear replies, "Yes, Killer Whale, but they also have lead encased within the concrete. You're more than right. They'll never pick up the signature until it is too late. OK, Stingray Charlie, you hotwire the motor and Stone Fish, you rig the shaped charge to go off when they try to ignite this bird."

Petty Officer Duke Moran, call name Stingray Charlie pipes back, "You got it, Papa Bear."

Ensign Dizzy Lee, call sign Stone Fish, mikes, "I'm on it, sir."

On the oil tanker, "La Senorita," Lieutenant Cody Custer, Chief Ryan Duffy, Chief Spider Sinclair and Master Chief Bull Sherman make their way to the modified cargo hold of the oil tanker.

Tiger Fox says over his COMM, "Papa Bear, the instructions were perfect. We're here and are going to work. All is quiet."

"Tiger Fox, copy that. We'll be done in ten minutes. Over."

Tiger Fox clicks his mike in recognition and affirmation and then says, "See you back at the ranch."

By 02:30 A.M, Tiger Fox has his men working on the motor and inserting the shaped charge into the missile. Suddenly, Cody Custer's SEAL team hears a match strike.

Simultaneously, over on the El Toro, Colt Jackson and his men are also working away---hotwiring the missile's ignition system. On this ship at the same time, the SEALs hear footsteps coming down the ladder to the cargo bay where they are working. Stone Fish and Stingray Charlie are bent over the missile, working on it. Stone Fish has the shaped charge in his hands.

Colt Jackson clicks his microphone and holds his fingers up to his lips. All of the SEALs freeze. They are invisible with their active cloaking camouflage engaged. The only problem is, the shaped charged Stone Fish is holding is not cloaked. To an outsider of the SEAL team, it appears to be floating in mid-air.

The Venezuelan Islamic State guard approaches the missile and looks at it, quizzically. "Que es? (What is this?). Como? (How is this possible?)"

Back over at the oil tanker La Senorita, the guard there is also making his rounds. The guards patrol every half-hour. This was their 02:30 patrol. He is casually puffing on his cigarette.

Tiger Fox's entire SEAL team holds its breath. They are invisible, but Black Widow---Chief Spider Sinclair, is working on the wires of the motor. As the Venezuelan Islamic guard looks at the missile, he sees the wires, seemingly floating in the air. The guard immediately draws his weapon. He is holding a Glock .45. One squeeze of that trigger and the whole ship will wake up.

The SEALs remain still and hold their breath. The guard approaches Black Widow, who has his back turned to him.

The guard says to himself, "Como es posible? (How is this possible?). Es una fantasma? (Is it a ghost?)"

The guard is walking directly toward Black Widow and walks right past Tiger Fox. As the guard strides by, only two feet away from Black Widow, Tiger Fox strikes.

Lieutenant Cody Custer swings a karate chop down with brutal force on the outreached, unsupported hand of the guard, which is carrying the Glock. The gun rattles to the floor and the guard gasps in horror. He believes he is being attacked by a ghost.

The shocked Venezuelan Islamic State guard, cries out, "Es una fantasma! FANTASMA!! (It is a ghost! A GHOST!!)." The guard swings wildly and makes physical contact with Tiger Fox. There is no going back now. Tiger Fox draws his Microtech DOC Flipper Knife out of its sheath and plunges it into the back of the neck of the native Venezuelan, who converted to the Jihad. The guard slumps over immediately, as his spinal cord is severed.

Lieutenant Cody Custer looks at his men, "Wrap it up, men. We've got to get moving. Someone will come looking for him soon when he doesn't show up. Make sure everything is put away and the missile looks exactly the same as when we arrived."

All the men nod in agreement and finish their tasks.

In the meantime an eerily similar scenario is playing out on Colt Jackson's ship. The guard is now off the ladder and is quickly approaching the shaped charge, which appears to be floating in the air. He is only six feet away from Stone Fish, staring intently at the floating object.

His eyes squint and his forehead turns into a frown, as the light of recognition dawns on his face. He can see it is C-4 in a shaped charge. He starts to pull his Glock from his holster, but Stingray Charlie jumps him. The guard is completely startled. He lets out a cry, "Ahhhh!"

In the meantime, Master Chief Jack Flintlock, has come up behind and grabbed the guard in a death grip. Killer Whale has his thick, powerful arm, wrapped around the guard's neck. Like a massive Python, Killer Whale squeezes the life out of his prey. The guard slumps over dead.

Colt Jackson orders them, "OK, men. We need to finish up. They will search this ship from stem to stern when the guard doesn't return from his rounds."

The men say, "We're on it."

Colt says, "Great move Stingray Charlie, great work, Killer Whale."

Stingray Charlie and Killer Whale respond, "Thanks, sir."

By 02:45AM, Eastern Time, the SEALs from both vessels finish their work and carry the bodies to their harpoon ropes. They throw them both overboard. The noise of the great ships covers any splash the bodies make in the deep, blue Caribbean Sea, below.

The SEALs repel down the ropes, retract the hooks from off the wall of their respective vessel and pull them down from the tankers. There is no evidence of any of them being there this evening. They jump onto their Zodiacs and speed toward their rendezvous point with the Raider. The mission is a complete success.

By 09:45AM, on the morning of October 6th, the SEALs are approaching the Naval Amphibious Base in Little Creek, Virginia. Daylight is not a friend to the U.S. Navy SEALs, particularly in this situation. Colt Jackson speaks to the pilot, Commander James Forrestal.

"Thanks, again, James. This bird is beautiful. Unfortunately, we're not going to be able to land on the ground with her in broad daylight, gear and all."

Commander James Forrestal replies, "It's been a pleasure, Commander. I'll let our friend know everything went OK."

"Please give Harry my best regards and say thanks a million!"

"Will do, Colt. You guys ready?"

"Without being flip, these guys were born ready! Tally hoe, see you on the other side, Commander Forrestal!"

The entire Navy SEAL team turns to Commander James Forrestal, salutes him and then all jump out the side door of the helicopter, which is 10,105 thousand feet in the air. The SEALs form a "star" formation on the way down.

If anyone sees them from the ground, below, they will look like just another run-of-the-mill training mission for the good, old U.S. Navy.

Colt Jackson speaks to his men over his COMM as the land rushes up to greet them, still far below.

"Great job men! I hate to be a real pain, but no games, today."

Master Chief Jack Flintlock is the first to complain, "Ahhh, come on, sir. Can't we have a little fun? Black Widow and I always like to be the last one to open our chutes---we always compete to see who will chicken out first!"

Chief Spider Sinclair supports his SEAL brother and friend, "Yeah, come on Commander, we won't push it past 500 feet, we promise!"

Colt speaks to them seriously, "Gentlemen, I love your sense of humor and competitive nature, but today, after this of all missions, we don't want to raise any suspicions. We'll all open up at 1,200 feet. That's still 800 feet below the legal limit for civilians. We don't want any injuries and we want to skedaddle out of here ASAP. OK?"

The SEALs all reply, "AYE, AYE, SIR!"

The SEALs open their parachutes, all at once. A little boy walking with his mother on a beach some distance away points his finger up to the sky. The eight parachutes are a wonder to the young man.

The little boy says to his mother, "Mommy? Look! I want to do that when I grow up!!"

The mother looks sternly at the young lad and says, "You should be a doctor, or a lawyer, instead, honey."

"But that's boring, Mom. I want to be a U.S. Navy SEAL, just like my dad!"

The mother inhales and exhales deeply. She walks down the beach with her little man, knowing one day in the future she'll be worrying about him, too.

By 07:00PM, Colt is back in his Georgetown Park condominium, as if nothing happened at all, the last 48 hours. Colt looks at the screen on his phone. His cell phone screen reads:

"43 Missed Calls"

Twenty-eight of them are from Lori Sanders. While he's reviewing the missed calls, the cell phone rings. Colt answers it.

"Hello?"

Lori Sanders is on the other end of the line, "Colt? I've been trying to reach you for DAYS!! ARE YOU ALL RIGHT? WHERE HAVE YOU BEEN? I've been so worried."

Colt pauses for fifteen seconds and is stone-cold silent on the phone. Colt is still furious with how Lori treated him during the meeting with the Admiral and General.

He says dryly, "I've been on vacation."

Lori answers, "Really? Where did you go? Your phone hasn't left your condominium. I got worried and came over to your condo complex and talked the security guard, Charlie, into buzzing me in through the main entrance to check on you. I walked through the complex, found your door and even knocked on it. You haven't been home."

Colt answers, "Frankly, Ms. Sanders, I didn't want to be bothered by you. I left the phone here, on purpose. In fact, I think we should just hang up. There is really nothing to talk about."

Lori urgently speaks to Colt, "Colt, I'm SO sorry. I should have never spoken to you that way. PLEASE forgive me."

"Look, I forgave you the minute after you talked down to me and treated me like a school child, or an idiot, you decide. But, either way, I'm afraid it really scared me. No one has ever spoken to me this way. Lori, I'm just going to hang up now. I don't want to slam it in your ear, but I have to really think about this. Have a good night, Ms. Sanders."

"Colt, PLEASE!"

"Goodbye." Colt hangs up the phone.

Lori Sanders lets the phone drop out of her hands and hangs her head, bursting out in tears.

There is a pile of used Kleenex sitting on her kitchen table, before her.

CHAPTER 19

EVERYTHING EXCEPT EL TIGRE

At the Pentagon, inside Admiral Steven Whistler's War Room, it is October 11th at 08:00AM.

Admiral Whistler and General Maxwell have a room full of people gathered in front of them. They are on the video phone with General Charles Jacobs. Lori Sanders is present, along with Brian Pierce and Kristi Kruger from the NSA. Some of the Admiral's top programmers and technologists are assembled in the room, as well.

If an attack occurs and the defenses fail, the country will need all hands on deck to recover. Lieutenant Izzie Brown, Lieutenant Jake Sampson and Lieutenant Ronnie Forrestal, the brother of the Raider pilot who helped Colt Jackson and his men, sit in the front row.

The President is in the Situation Room with the Vice President, hooked in by video conference.

Admiral Steven Whistler is looking at the image of General Charles Jacobs who is in his secure, NORAD facility in Colorado Springs. "Hello, General. We see lots of clouds this morning. It looks like you've been busy."

"We have. We flew all night out of both Altus Air Force Base, in southwestern Oklahoma and Marana Air Park, out of Tucson. We covered the nation. The planes just returned and are all back in their hangars. We're ready. All systems are a 'GO.'"

General Maxwell speaks, "Each ship has its own schedule for firing. It's up to the ground crew to decide when they launch the missiles. The Islamic State wants to inflict pure terror and randomly set off the nukes."

Lori Sanders contributes, "We believe the ship on the East Coast and the one in the Gulf, will fire around 10:00AM Eastern Time. This will be 09:00AM in Texas and Louisiana. Everyone will be at work and it will send a shockwave across the Mid-West and East Coast. The West Coast attack will occur a few hours later, about noon, Eastern Time, when all businesses are open.

President John Diaz asks a question from the Situation Room in the White House, "What about the ships out of Amsterdam?"

The Admiral says, "Right about now, sir, they are targeted by the USS Minnesota, under Commander Brian Tenclaw, call name 'Neptune.' He's a good man. He'll get them sir, before they light up."

Inside the USS Minnesota Common Submarine Radio Room, the "CSRR," it is 08:10AM Eastern Time and 03:10PM Qatar Time. Commander Brian Tenclaw has locked in the coordinates for the three tankers. They are sitting off the East Coast of the United States, just outside of the 12-mile international water mark.

Commander Brian Tenclaw orders his torpedo crewman, "On my mark, three, two, one...FIRE ONE! FIRE TWO! FIRE THREE!"

The U.S. Navy submariner at the controls of the smart torpedoes, does his job. The three deadly fish are on their way to three different cargo vessels. The torpedoes are extremely precise, due to their advanced guidance systems. The USS Minnesota nuclear attack submarine has marked the precise location on each ship where the missiles are located. The enormous blast created from these 19 feet, 3,695 pound explosive fish, will sink a simple cargo ship in minutes.

The USS Minnesota torpedo crewman responds, "Aye, aye, sir!"

Commander Brian Tenclaw talks into the encrypted video COMM channel, plugging him live into the Situation Room, Cheyenne Mountain at NORAD, as well as Admiral Whistler's War Room. "The fish are away. You can jam the ships' communication capabilities, now."

Lori Sanders turns to Brian Pierce and commands, "Brian, jam them, as planned."

Brian Pierce dutifully answers, "Yes, ma'am." Brian flicks his fingers across his key board and the three ships are immediately enveloped with a blanket of electronic haze, which blocks all communication to and from the ships.

Moments later, three explosions are clearly heard over the video and audio feed coming from the USS Minnesota.

Commander Brian Tenclaw confirms over the COMM channel, "Three hits. I repeat…three direct hits. The ships are already listing. Send for survivors. Good luck, everyone. It's time for us to run silent and run deep. Neptune and the USS Minnesota---OVER AND OUT."

Admiral Whistler heaps praise on Commander Tenclaw, "Great job, Neptune. I owe you a beer. See you soon. Over and out."

The screen from the USS Minnesota nuclear attack submarine goes dark. The sub and its crew have done their job and sunk three immediate threats to the East Coast of America.

At the Royal Palace in Qatar, in the Royal salon of the Emir, it is now 04:00PM. Emir Sheikh Muhammad Al Therani, Caliph Abu-Bakr and Admiral Muhammad Mubarak are gathered together. They are watching the American version of CNN, waiting for the station to stop airing when the EMP attack strikes. They have been fixated on the channel for the last few hours. They are growing anxious.

The Emir Sheikh asks the question, "The LUBEROL ships should launch any second now. Can we contact them and see how they are doing?"

Admiral Muhammad Mubarak responds, "Of course, your majesty. I will call QEMCO and have them use their code words to see if everything is on track. They have direct communication capability with LUBEROL's fleet."

Caliph Abu-Bakr concurs, "This is a very good idea. Let's make sure everything is on track."

Admiral Mubarak speaks in Arabic to QEMCO. He waits for them to come back to him.

Admiral Mubarak tells the two men, "QEMCO cannot reach them. Maybe they already launched?"

The Caliph answers, "Impossible. This is the American-based version of CNN, picked up from satellite, right?"

The Emir says, "Yes, as we planned."

The Caliph states the obvious, "CNN broadcasts from the Time Warner Center in New York. Any blast from the ships would decimate that city and obliterate CNN's broadcasting capability."

Admiral Mubarak responds, "Of course, you are right, Caliph. Let's give them a little more time. In any event, the El Toro and La Senorita should be just about ready to strike also, it's 10:00AM on the East Coast of America. This will be a great day for all of us. Allahu Akbar!"

Caliph Abu-Bakr and Emir Sheikh Muhammad Al Therani respond, "Allahu Akbar!"

Inside the Pentagon in Admiral Steven Whistler's War Room, it is now 10:02AM Eastern Time. Lori Sanders receives a call on her NSA PRISM phone. She listens intently and looks around the room. General Shane Maxwell looks at her, questioning her with a "What's up?" look.

"General? Admiral? I just received a flash report of small explosions---right off the coasts of both New Orleans and Charleston."

They all look at each other, thinking on the fly.

Admiral Steven Whistler immediately growls, "Does anyone know where *Colt Jackson* is?"

Lori answers, "I spoke to him in his Georgetown condo, over the phone, a few days ago. He said he was on vacation."

The Admiral replies, "Yeah, right. That man doesn't take vacations."

General Maxwell turns to the Admiral and wonders, "Admiral, do you think..."

Admiral Steven Whistler turns to Lori and says, "Ms. Sanders, if you could please deploy, with the General's permission, the considerable resources of the NSA to finding Commander Colt Jackson, our country would be EXTREMELY grateful."

President Diaz jumps in, "What's Colt doing now, Admiral? What's going on?"

Admiral Whistler responds to the President's query, "We're not sure, sir. But, we're going to find out. With the explosions, we have to investigate immediately, regardless. The world knows we track every single vessel approaching our shoreline. I'll send the Coast Guard out immediately to offer support and 'assistance.' I'll also tell them to come well-armed and prepared. Those ships will have military guards from the Islamic State on them, that's for sure."

General Maxwell orders Lori, "Ms. Sanders, redirect the jamming frequencies for the Gulf and the area outside Charleston on the GPS coordinates of those ships, immediately."

"Yes, sir." Lori turns to Brian Pierce. "Brian, make it happen, NOW."

"Yes, ma'am." Brian's hands are flying over the keyboard. Kristi Kruger is watching amazed at how fast Brian changed gears and modified the jamming capabilities of the United States Government.

On October 11 at 10:45AM in Charleston, South Carolina, the U.S. Coast Guard Cutter, "Mohawk," WMEC-913, is equipped with a Melara 76mm Gun and two Browning .50 Caliber Machine Guns. It approaches the oil tanker, "La Senorita."

The Coast Guard Captain screams over the loud speaker, "Ahoy there, La Senorita! Prepare to be boarded. We heard an explosion. Is everything all right?"

The vessel La Senorita, responds in a distinctly Spanish accent, "Everything is AOK, here. One of the crew lit a cigarette and a gas can blew up. Everything is under control now, though."

The Coast Guard Captain responds, "Very well. La Senorita, prepare to be boarded, regardless. If we are fired upon your ship will be sunk, is that clear? This is a national security investigation. You are ordered to STAND DOWN! AM I CLEAR?!"

The oil tanker La Senorita meekly answers, "Si, Senor."

A similar scene is playing out in the Gulf of Mexico, a thousand miles to the west. Here, the El Toro is being boarded by the U.S. Coast Guard Cutter "Bear," WMEC-901.

Captains of both Coast Guard Cutters bring along their high-definition, wireless video cameras.

Inside the Pentagon, in Admiral Steven Whistler's War Room, Lori Sanders has been busy giving orders to Kristi Kruger. Kristi has directed the satellite feeds, tracking the El Toro and the La Senorita, onto separate screens in the Admiral's War Room. She directs her to go live with both Captains' feeds on two additional screens. Everyone watches the boarding of the vessels.

Lori Sanders instructs Kristi, "Kristi, bring up the feeds from the satellites, focused on the two vessels. When you have that complete, go live with the two video cameras of the Captains."

"Yes, ma'am."

The room stares intently at the wireless camera feeds both Captains are transmitting back to the War room, Cheyenne Mountain and the White House. Both feeds show the Captains approaching a stairway. They are accompanied by a platoon of armed men from the U.S. Coast Guard. As they enter the specially-built cargo holds, the North Korean SA-2 missiles become clearly visible on the screen. Several gasps come up from the War Room. The Captains, and their men, on each boat, also spot the Islamic guards."

The Captain of USCGC Bear orders the guards on the El Toro, "Drop your weapons! NOW! By order of the U.S. Coast Guard."

One of the Islamic State guards screams in defiance:

ALLAHU AKBAR!

The guard then opens fire with his Uzi submachine gun on the Captain and his men. Two Coast Guard officers drop. The rest of the squad opens fire and kills every guard in sight.

On the oil tanker La Senorita, the Islamic guards do not wait; they open fire, hitting the Captain of the Mohawk in the thigh. He cries out "Ughhhh!"

The Captain's men pull him back up on the deck. They are equipped with grenades, but have been briefed on the risk of spreading collateral radiation damage if the nuclear weapon on the SA-2 is breached. They will have to do this the hard way.

The Captain turns to his Lieutenant and says, "Take command. Open the roof of the cargo hold and set up firing positions. We take the high road and we'll shoot them like sitting ducks. Do it, NOW!"

The Lieutenant of the USCGC Mohawk answers his Captain, "Sir, yes sir!"

The squad of men opens the cargo bay's roof. It reveals the North Korean SA-2, clearly visible now through the Captain's wireless camera and through the Keyhole satellite feed. The War Room in the Pentagon can clearly see motor parts blown up behind the missile. There has been an explosion, which obviously crippled the SA-2, preventing its launch.

Admiral Steven Whistler takes in the scene, which shows remarkable detail on their HD screens. "Take a look at that!"

He points to the damage. Then, speaking into the COMM line with the Lieutenant who is now wearing the wireless head cam and investigating the "La Senorita," in Charleston, says, "Lieutenant, if you can please give us a good view of the rocket motor area behind the missile."

The video camera pans to the area where the ignition of the engine is supposed to occur on the SA-2, at the bottom of the rocket.

The Lieutenant explains what he sees, "There has been a small explosion, sir. Not enough to damage the nuke, because it looks like it's in perfect condition."

He pans up to the top of the missile, revealing the perfect outline of the missile, unscathed from the explosion.

General Shane Maxwell, also taking in the scene from both HD feeds from the ships, says to the Admiral, "Those explosions look like they are identical, don't they?"

Lori Sanders says out loud what everyone was thinking. "Did Colt Jackson do this?"

Admiral Steven Whistler replies, gruffly, "He'll be in the brig, forever, if he had anything to do with this. Did you find him, Ms. Sanders?"

Lori Sanders answers, "Apparently, he's in his condo in Georgetown, on vacation."

Admiral Whistler glares, "Get his butt in here, immediately, on my orders."

Lori answers "Yes, sir."

CHAPTER 20

DISASTER STRIKES

Offshore of Long Beach, California, on the oil tanker "El Tigre," it is 9:32AM Pacific Time, 12:32PM Eastern Time and 07:32PM Qatar time on October 11th.

The Islamic guards on the El Tigre take out their prayer rugs and pray to Allah one last time, before they throw the switch which will help destroy America. The lead guard finishes praying and stands up. The other guards stand, in turn. All of them raise their weapons.

The Islamic State guards all chant,

"ALLAHU AKBAR!
ALLAHU AKBAR!
ALLAHU AKBAR!"

The lead guard yells out and waves his rifle toward the sky:

القادمـــة الأيــــام الله صــــلى

Allah ybarek feek al-ayyam alqadima!

(May Allah bless the days to come).

On October 11th at precisely 09:32:50 Pacific Time, 12:32:50 Eastern Time, 07:32:50 Qatar time, the lead guard presses the ignition button of the SA-2, from the protected control room. The missile leaps off the launch pad and fills the cargo hold with fire and smoke.

As it arcs through the sky, the SA-2 ultimately reaches a top speed of Mach 3.5, or 2,664.23 miles per hour.

Inside the Pentagon in the War Room, Lori Sanders screams, "GENERAL, ADMIRAL! We have a confirmed launch from the El Tigre oil tanker off the coast of Long Beach."

Admiral Steven Whistler looks at General Charles Jacobs and states what both men are wishing, "Angel Shield better work!"

SA-2 Missile

Public domain image.

Inside Cheyenne Mountain at NORTHCOM Command Center, in Colorado Springs, Colorado, General Charles Jacobs speaks out, confidently, over the video conference line, "We are 100% sure. Watch the magic of the Angel Shield at work."

President Diaz tells the Situation Room "Hold your breath, people."

Meanwhile, all around Southern California, it is business as usual:

At the Santa Monica Pier, Santa Monica, California, it is Yom Kippur on October 11th at 09:33AM Pacific Time, 12:33PM Eastern Time, 07:33PM Qatar Time. The Ferris wheel on the pier has just cranked up for the day. It is warm in Southern California, already in the high seventies.

Los Angeles International Airport, LAX, is one of the busiest airports in the world. Today it is also very busy, as usual. Planes are already stacked up, waiting their turn to come in for a landing.

At Disneyland in Anaheim, its gates have been opened since 08:00AM and the park is quite busy for a fall day. The sunshine and warmth have brought droves of people out.

At the Lockheed Martin Skunk Works, in Palmdale, workers are busy at work on some of the most advanced, top secret aircraft designs of the future.

Downtown in Los Angeles, in the financial center, it is "same old, same old." Cars are honking, with thousands of people walking in and out of tall buildings.

On the 405 Freeway, there is an enormous traffic jam, with cars backed up for miles.

In Beverly Hills, Rodeo Drive is being cruised by Rolls Royce, Mercedes Benz and BMW motor vehicles. Its sidewalks are crowded with people.

At UCLA, The University of California at Los Angeles is crawling with a very active student body. Everyone is heading to a class.

At Santa Monica High School a computer science teacher is teaching the class. Everyone is looking at an iPad.
At Camp Pendleton Marine Corps Base, near San Diego, a drill Sergeant is screaming at his recruits.

At the border crossing in Tijuana, Mexico and San Diego, California, the line of cars to get into the United States is already a mile long.

With Southern California one of the busiest places on earth, it also is the center for hundreds of billions of dollars of productivity. It creates jobs for tens of millions of people, worldwide.

Burbank is the headquarters for Walt Disney Company. Walt Disney Companies generates over $45 Billion per year in revenues.

El Segundo is the headquarters for DirecTV. DirecTV is the world's largest satellite television company. It generates over $31 Billion in revenue per year.

Santa Ana is the headquarters for Ingram Micro. Ingram Micro is the biggest distributor of technology. They have revenues of over $42 Billion per year.

San Diego is the headquarters for Sempra Energy. This $10.6 Billion power company is busy counting its money, billing its customers throughout Southern California.

San Diego is also the headquarters for Qualcomm. Qualcomm is the world's largest manufacturer of semiconductors for cell phones. It generates over $24 Billion in revenue per year.

Irvine is the headquarters for Western Digital. Western Digital is the world's largest manufacturer of hard drives for computers. It generates over $15 Billion in revenue per year.

Thousand Oaks is the headquarters for Amgen. Amgen is one of the world's largest bio-tech companies. It generates revenues of over $18 Billion per year.

The San Onofre Nuclear Power plant is located right in between San Diego and Los Angeles. It sits on the steps of Camp Pendleton. It is operated by Southern California Edison and generates electricity for Edison International and Southern California Edison. One minor issue is that it also was built on the San Andreas Fault.

On this day, October 11th at 9:33:15AM Pacific Time, 12:33:15AM Eastern Time and 07:33:15 Qatar Time, at the top of its trajectory, nearly twenty miles above Los Angeles, the 10-kiloton nuclear weapon detonates within 25 seconds of lift off.

At high altitudes, there is less air mass to absorb radiation and convert it into a blast wave. The lower air density at this altitude, on the other hand, allows tremendous, highly-charged radiation energy to be released as ionizing gamma radiation and x-rays.

This is the trade-off of an EMP attack, versus a ground-level detonation of a nuclear device. EMP attacks create far broader infrastructure damage and cause much greater economic damage, especially for sophisticated economic societies. They pose a mortal threat to nearly all nations on Earth.

In arguably the most sophisticated geographic location, electronically speaking in the United States, society will now feel the full force and effect of a perfectly executed EMP attack.

Los Angeles and Southern California, particularly, will never be the same.

Traveling at the speed of light, the bright, double-flash of the explosion causes people to look up from all over the region. Many become temporarily blinded and nearly everyone cowers when they hear the thundering power of the explosion. It comes just seconds after the great flash of light.

In an instant, all commerce comes screeching to a halt. The electric grid is dead. All communications stop and most vehicles come to a halt in the radiating area of the detonation. Its highly-charged ions fry nearly everything electronic and most cars simply stop.

Some vehicles keep moving, however, because their larger tire structures provide enough insulation and grounding for the car---turning it into a virtual Faraday cage. A Faraday cage causes highly-charged ions to stream around an object and short out through a ground, thus protecting electronics which may be inside of the "cage."

Unfortunately, only a few cars survive. All "non-hardened" businesses, which depend on electricity, grind to a halt instantaneously.

Critically, every single electric utility is destroyed, from within. The higher-voltage transformers are permanently damaged by the highly-charged radiation energy which invisibly invades them. They cause extremely high voltage to stream into the conductors of the power plants. They completely fry out.

These major transformers are not replaceable. The manufacturing process for them is extremely complicated and *each* transformer can take up to one year to build.

Further complicating matters, manufacturing for them is done primarily in Germany and China, where the units are produced serially. This means building a "Henry Ford" production line for them is impossible.

Now Southern California needs to replace at least FIFTY of these substation transformers.

In the meantime, the San Onofre nuclear plant begins to heat up. Its wiring and electric system are fried. A full-scale nuclear meltdown appears inevitable.

Cell phones don't work, home phones don't work, televisions and radio don't work.

What is working are the hardened electronic operations of the defense industry. Twenty-seven stories underground at Lockheed, business, as usual, prevails. The CEO of Lockheed picks up the hot line and is buzzing Admiral Steven Whistler on the hotline.

It is now 12:37PM Eastern Time. Admiral Steven Whistler is still in the War Room and sees his hotline ringing, with the caller ID indicating

Lockheed Martin is on the phone. This line is only used in national defense emergencies, and is hardened against an EMP attack.

"This is Admiral Whistler. Who is calling?"

Inside the Lockheed Martin Skunk Works, twenty-seven stories underground, in Palmdale, Catherine Jepson is on the line. "Hello, Admiral. This is Catherine Jepson, CEO of Lockheed Martin. I thought you would appreciate a sit-rep. We're in touch with our other defense industry partners and everything below ground is humming along just fine. However, the situation above ground is a complete disaster. The electric grid is completely taken out."

Admiral Steven Whistler is in shock. He looks around incredulously and then looks directly at General Charles Jacobs.

"Can you repeat that? A Disaster?"

"Yes, sir. The nuke went off and the ionization came down across all of Southern California like rainfall. Angel Shield did not work. Repeat Angel Shield did not work. It is a bit strange, however, because everything above Bakersfield appears to be OK. I really don't understand it, sir."

"How's that, Ms. Jepson."

"Sir, I did my graduate degree in nuclear engineering and weapons systems. An EMP with a 10-kiloton blast, which by our measurements is about the size of the blast, should have wiped out San Francisco, too. But, it certainly should have destroyed much more of California. It's like Angel Shield worked up there, but down here, we got screwed. Excuse the expression, sir."

"Thank you for that insight, Ms. Jepson. I'm sure the President will weigh in on this and do everything in his power to bring assistance and electricity ASAP. Hang in there, Catherine."

"Yes, sir, we will." She hangs up.

Admiral Whistler looks at the video feed and says, "General Jacobs, can you explain what happened here?"

At NORTHCOM Command Center, in Cheyenne Mountain, General

Charles Jacobs responds, “I don't understand it, sir.”

Admiral Whistler tells him, “I'm deploying armed forces to both Altus Air Force Base, in southwestern Oklahoma and Marana Air Park, in Tucson. All planes are grounded, if they can even fly at all in Arizona. Is that understood? I am going to have the Joint Chiefs issue an order to the U.S. Army to post a 24-hour guard around those planes. We're going to check every single one of them and get to the bottom of this complete and total disaster!”

General Charles Jacobs replies, “Yes, sir! I'll see to it.”

At 12:42PM Eastern Time, Admiral Whistler yells at Lori Sanders, “GET ME Colt Jackson. IN HERE IN THE WAR ROOM, ON THE DOUBLE!”

Lori answers, “He's on his way, sir.”

Around the room, people are looking at the images of Southern California from the Keyhole satellite feeds. Hundreds of thousands of cars are just stopped on the freeways. Within a couple of minutes of the explosion, there were nearly fifty other ground-based explosions.

The people in the room watched in horror as the Keyhole satellite projected live images of large, commercial aircraft falling from the skies. Many of them crashed into the Pacific Ocean, as the brave captains of the planes tried to ditch them into the water---away from population zones. Thousands of souls perished, right before their eyes.

The President is furious and he is raging. Fuming mad, he screams, “General Jacobs! HOW IN THE HELL IS THIS POSSIBLE!! You told us you were 100% confident in the system! I'm going to skin you alive for this, if you are to blame. Do you understand me?”

General Jacobs replies, “Yes, sir. I can assure you sir, Angel Shield works. It had to work because significant parts of California were not destroyed, that otherwise should have been.”

The President states mockingly, over the video conference, “This is what you call working? This is outrageous!”

The President is raving, out-of-control, and is acting very un-presidential at this moment.

Lori Sanders speaks directly to the Admiral, “Sir, Commander Jackson is on the premises.”

President Diaz demands, “What does he have to do with this? He wasn't even involved with this operation and had nothing to do with Angel Shield?”

Admiral Whistler replies, “Sir, I need to speak with Commander Jackson and ask him some questions.”

The President, in shock from the attack on L.A. replies, “O.K. Admiral, but we need answers NOW!”

“Understood, sir.”

At 12:53PM Eastern Time, Colt Jackson enters the War Room at the Pentagon in his blue jeans and black Hugo Boss, V-neck t-shirt---the same outfit he wore on his date with Lori Sanders in Denver.

Lori's eyes meet Colt's but he looks past her to the satellite feed streaming in from L.A. Colt knows instantly what happened. He also knows this is not an “I told you so moment.”

Colt Jackson walks directly to Admiral Whistler, “Reporting as ordered, sir.”

Admiral Steven Whistler is spitting venom:

> “NO BULLSHIT, COMMANDER. DID YOU, OR DID YOU NOT, DISOBEY A DIRECT ORDER AND ELIMINATE THE THREAT FROM THE EL TORO AND LA SENORITA OIL TANKERS?”

Lori Sanders and the entire room, including the President of the United States over the video conference, sit in silence, waiting with apprehension for Commander Colt Jackson's response.

Colt Jackson keeps his cobalt-blue, steel eyes locked on the Admiral with a fierce stare which would make anyone fear. Colt Jackson remains silent. Admiral Steven Whistler bellows, “Commander JACKSON, I am ordering you to answer. RIGHT NOW, MISTER!"

Colt Jackson's cold, blue-steel gaze remains locked on the Admiral. Then, he slowly turns it directly upon the eyes of General Shane Maxwell and finally, turns and looks intently, into the eyes of Lori Sanders. Lori nearly faints. General Maxwell starts to sweat. The Admiral presses on.

Admiral Steven Whistler shouts, "LAST CHANCE, COMMANDER. DID YOU HAVE ANYTHING TO DO WITH THOSE EXPLOSIONS ON THOSE SHIPS?"

Colt Jackson refuses to reply, but his gaze is one of resentment and contempt.

Admiral Steven Whistler barks out an order, "MPs! Escort Commander Jackson immediately, to the brig. You're going to rot in there until we have answers. We're going to get to the bottom of this, but right now we have a national emergency to attend to. Take him away! MPs, escort the prisoner out of my site."

The MPs reply, "Sir, yes sir!"

They escort Colt Jackson out of the Admiral's War Room. He locks his glare on Lori Sanders as he is taken off in nylon-white, double-loop handcuffs. Lori Sanders has tears in her eyes.

At Altus Air Force Base, Oklahoma and Marana Air Park in Tucson, Arizona, with one phone call the Chemtrail-fleet bases are locked down.

In Altus, the fleet is secured by U.S. Air Force airmen. In Tucson, the U.S. Army's Fort Huachuca sends dozens of personnel to lock down the airstrip and hangars where the large-bodied, white aircraft of the Chemtrail fleet lie at rest.

No one is allowed in, or out, of the aircraft. Forensic, chemical specialists are on their way from Fort Detrick, Maryland to inspect every inch of the planes' spraying systems. The Admiral is determined to find out exactly what happened to Angel Shield.

Back at the Pentagon, Admiral Whistler speaks over the video COMM line, "General Jacobs, the FAA has the flight path for each Angel Shield aircraft. Can you please send me the base logs, so I can compare them for any possible discrepancies? I've already received the FAA log. It was relatively easy for them to track the aircraft, since they were all flying in the early morning hours of the 11th. There isn't much traffic then, and their crisscross patterns make it easy to track them on radar."

General Charles Jacobs feels the noose tightening. He swallows. "Yes, Admiral. You'll have it in minutes. It should corroborate 100% with the data from the FAA." While on the video conference, the General turns to his adjutant and says, "Send the Admiral and General Maxwell the log book, immediately."

Admiral Whistler replies, "Thank you, General. We'll be in touch, shortly. Don't go anywhere."

General Charles Jacobs decides to alert the War Room and everyone else, "Sir, I need to bring up other threats---like the North Korean satellite. What are the orders from the Joint Chiefs?"

The Admiral forcibly responds, "If it even moves a muscle, turn it into a space rock. Do not hesitate; do not worry about blowing our secret with the Chinese, or the Russians. Is that clear?"

General Charles Jacobs answers, "Crystal, sir."

General Shane Maxwell turns to Lori. "I want you and your team to get to Ft. Meade and the Glass House immediately. Have every inch of the ocean route of those three tankers, El Tigre, El Toro and La Senorita, checked out. I particularly want to know what happened to the El Toro and La Senorita. I don't care if you have to stay awake for a week. Get back to us IMMEDIATELY with any Intel."

Lori Sanders replies, "Yes, sir." She turns to Kristi Kruger and Brian Pierce, "Let's go!"

They exit the war room and head to NSA headquarters, about a half-hour away. They are given a Pentagon escort and they are flying down the Baltimore-Washington Parkway.

In the Royal Salon of The Emir, in Qatar it is 08:10PM, 01:10PM Eastern Time and 10:10AM Pacific Time.

Emir Sheikh Muhammad Al Therani, Caliph Abu-Bakr and Admiral

Muhammad Mubarak are watching CNN on the 110-inch Samsung Ultra HD television in the Sheikh's Salon. Everyone is growing frustrated with the lack of news and inability to communicate with the ships.

Caliph Abu-Bakr fumes, "I don't like the looks of this, my brothers."

Emir Sheikh Muhammad Al Therani states, looking seriously, "Admiral, what do you have to say for yourself? You were responsible for this aspect of the operation in the Atlantic and coordinating with our brother, General Jose Lopez in Caracas. What happened?"

The Admiral answers in confusion, "My Emir, my great Caliph. I do not understand. The missiles were ready for launch when I visited Caracas and met the General, personally on September 26th, at the top of Waraira Repano Park, above Caracas. I am certain we were not followed."

Suddenly, from the 110-inch Samsung Ultra HD TV, there is breaking news from CNN.

"We interrupt this broadcast with a national emergency message from the President of the United States."

Inside the White House in the Oval Office it is 01:11PM Eastern Time, 08:11PM Qatar Time and 10:11AM Pacific Time, October 11th.

President John Diaz looks into the camera. His looks could kill…

> "My fellow Americans and citizens of the world. Today, I have the sad duty to inform you the city of Los Angeles and the entire Southern California region has been subject to a nuclear, electromagnetic pulse attack. Before everyone panics, this is not a Hiroshima-type of attack.
>
> There was not a massive fireball on the ground, creating millions of deaths. This was an EMP attack. For those not familiar with it, it destroys the electronic infrastructure of pretty much anything it touches. The corridor between San Diego, Los Angeles and Santa Barbara, is one of the most electronically-wired areas of the country. It also has the most automobiles.
>
> As of this moment, there is no electricity, there are very

few working automobiles, there are no television broadcasts and there are major potential problems, including the possible meltdown of the San Onofre nuclear plant. If you have loved ones in this area, do not try and phone them. Communications systems are down.

Your best bet is to use the Internet to send your loved ones an email message. The Internet is a living web and was designed by the Defense Advanced Research Project Agency—DARPA, to withstand a nuclear attack. I'm not going to sugar-coat it. This is a massive blow to the United States of America and we have closed all exchanges and trading. No one will get rich off of this cowardly attack, this I can promise you. Instead, we will fight back.

We will get to the bottom of who is responsible for this attack and destroy them. COMPLETELY. We will handle this with resolute force.

I will come back to you, as a nation, with our plan to aid and assist Southern California repair itself from this devastating attack. Initially, I am asking every power company in America to send two transformers to the Department of Homeland Security. These, along, with some high-capacity transformers the United States has kept buried in the limestone caves of Missouri, should get California back on its feet, immediately.

In the meantime, I am resending Congress the Shield Act bill, quadrupling the budget to harden every inch of electronics throughout this great nation. For everyone in California who has access to a hardened electronic wire, and anyone receiving this broadcast over the Internet, spread the word to NOT PANIC.

Help and resources are on the way. Water, food and portable generators have already been shipped out of Homeland Security storage facilities in Arizona. They'll be arriving in Los Angeles and San Diego, within hours. That is all. God bless this country and God bless each and every one of you."

The President ends his communication.

Back at the Royal Salon of The Emir at 08:14PM Qatar time, 01:14PM Eastern Time, 10:14AM Pacific Time, the Emir, Caliph Abu-Bakr and the Admiral erupt in happiness. They kiss each other on the cheeks and slap each other on the backs.

Caliph Abu-Bakr turns to the Admiral, "My congratulations, Admiral. We must congratulate the General. He has proven to everyone now, he deserves to be crowned the Caliph of ISSA---the Islamic State of South America!"

The Emir agrees, "I knew I was right about him. The gift of the six virgins, including his prized Fatima, must have given him strength. I am proud of him."

The Admiral concurs, "Indeed, he is a fine and good man. He will have some explaining to do with the Americans. I hope Venezuela survives what may be coming upon them! We are hopeful the brave Muslim warriors on the tanker can be blamed, instead of the government. This would be the best sacrifice they can make. They know this, as well. If they are caught, they will say they are committed Venezuelans against the Catholic Church---Allahu Akbar! This will give the Americans the evidence of a splinter group cell operating within Venezuela and not an organized attack by the country, itself."

The Emir Sheikh congratulates the Admiral, "Brilliant, Admiral. We must protect General Jose Lopez!"

The Caliph says, "Admiral, I agree with this. I do have one question, though. What happened to the other five ships?"

The Admiral looks at both the Emir and the Caliph and says, "I do not know. LUBEROL cannot reach their three cargo vessels. They are not communicating."

Suddenly, Admiral Muhammad Mubarak's phone rings. Caller ID indicates it is the CEO of QEMCO, Sheikh Sultan Bin Mohamed Rashid is on the line.

The Admiral answers, "Yes, my Sheikh, what is the news?"

Sheikh Sultan Bin Mohamed Rashid replies, "Admiral, I just got off the phone with LUBEROL in Amsterdam. The Captain of one of the cargo vessels was picked up by a local fishing boat. The other crewmen were picked up by the U.S. Navy and U.S. Coast Guard. THE SHIPS WERE TORPEDOED!"

Admiral Mubarak is stunned by this news. "WHAT? TORPEDOED? ALL OF THEM?"

The CEO Sheikh responds, "Yes, the captain said all of them were hit. He was able to watch them all sink into the Atlantic. A small fishing trawler brought him back to Montauk, at the very end of Long Island in New York. This is why it took him so long to make a phone call. There was no way to communicate what happened until now."

The Admiral states the obvious for the Islamic State, "This is bad news. The North Koreans will not launch the warhead from the Unha-3, their 'Bright Star' satellite, until considerably more damage is done to the United States. To provide a knock-out punch, we need this to happen. Bright Star flies directly over the U.S. and both poles, at an altitude of 500 kilometers. Firing a missile from this platform and detonating it at an altitude of only 100 kilometers, would create a great HEMP---a High Altitude Electromagnetic Pulse attack. This would blanket the ENTIRE country. There aren't enough transformers in the world to save them, if this were to occur."

The Caliph is thinking. He asks a question, "Do you think the North Koreans would strike if the strikes over New Orleans and Charleston are successful?"

Admiral Mubarak answers, "I would say the chances are over 80% the North Koreans would launch the missile from Bright Star. They hate the Americans and if they can help destroy the country and know they will survive a response, their leader will go for the jugular."

Emir Sheikh Muhammad Al Therani looks to the heavens above and says:

أكبر الله .استعداد على هو الله كان إذا

"Insha'Allah. Allahu Akbar!"

(If Allah is willing. Allah is great!).

CHAPTER 21

KEYHOLE & INTELLIGENCE ANALYSIS

Offshore at Ft. Meade inside the NSA at 02:05PM Eastern Time, October 11th Lori Sanders, Brian Pierce and Kristi Kruger are looking at the virtual holographic display of the two tankers. The displays are crystal clear and the Keyhole system can read a dime off of the deck of either the El Toro, or the La Senorita. The analysts are fast forwarding the footage, but the computer software filters don't pick up anything unusual during their journey between Maracaibo and their respective destinations---the ports of New Orleans and Charleston.

Brian Pierce is discouraged, "The computer's not picking up anything, Lori."

Kristi Kruger joins in, "Both ships' profiles don't change. There isn't an attack on them. It's like the rocket motors just went hay wire."

Lori Sanders is deep in thought. "Something had to have happened. I don't believe in coincidences. Those explosions were IDENTICAL."

Brian offers up a possible answer, "Maybe they had a betrayal within their own ranks?"

Kristi looks over, impressed with Brian's ability to always create alternative theories, to complex problems. "That's a good idea, Brian."

Kristi smiles at him. Brian's day has just been made. He smiles widely at Kristi and feels a flush go through his body.

Lori Sanders looks at Brian. "That's certainly a possibility. Given the similar damage patterns, whoever did this was well trained and trained by the same person."

Lori pauses and is deep in thought, again. She continues. "OK, this is a real possibility. But, before we can go with it, we need to do one more thing."

Brian asks, "What's that, Lori?"

Lori Sanders says, "We need to go over the journeys of both ships, one more time."

Kristi complains. "What good will that do Lori? I mean, with all due respect, we have the Cray's computing over the filters on the ships. Anything unusual would get flagged. I'm sure of it."

Lori responds, "I know the Cray Research Supercomputers are crunching this algorithm. That isn't the point."

Brian asks, "So, what *is* the point?"

Lori Sanders answers both their questioning faces. "The point is *our assumptions.*"

Kristi raises her eye brow and shrugs her shoulders, "You've lost me, Lori."

Brian is more direct, "I feel like an idiot. What did we miss? We've checked every inch of the trail between Maracaibo and their destinations. What did we miss?"

Lori Sanders revealing, finally what her hunch is, "This time, run it in *infrared.*"

The dawn of recognition lights up the faces of Kristi Kruger and Brian Pierce. "YES, MA'AM! WE'RE ON IT!!!"

In the Pentagon at Admiral Whistler's War Room it is 03:00PM Eastern Time on October 11th.

Admiral Steven Whistler is on the phone, speaking with some of the forensic chemists from Ft. Detrick who were whisked, via F/A-18F Super Hornet jet fighters, to Altus Air Force Base in Southwestern Oklahoma.

Ft. Detrick is a United States Medical Command, which used to be the center for the US biological weapons program. Given the national emergency, the double-seated F/A-18Fs were sent to pick up the chemists and scientists who would be dissecting the Angel Shield aircraft exhaust systems.

Ft. Detrick didn't have a runway, but the command closed down Military Road, which runs in front of the facility. It is several thousand feet long and the Super Hornets landed there and picked up their passengers. The fighter jets fly at a speed of Mach 1.8, or 1,190 miles per hour. Altus is 1,400 miles away and Marana Air Park is about 2,300 miles away.

The group in Altus has already been working in Oklahoma for an hour and the group which flew to Arizona, is just landing.

Admiral Whistler asks, "Dr. Fleckstein…What have you learned?"

Dr. Steven Fleckstein, PhD, responds, "Admiral, based on our preliminary analysis, these aircraft were not carrying metallic aerosols."

Admiral Whistler asks for clarification. "What do you mean, they weren't carrying metallic aerosols?"

Dr. Fleckstein states, "What I mean sir, is there is absolutely no trace of aluminum-oxide in any of the systems."

"What, pray tell, IS in it?" The Admiral asks.

"It appears to be a combination of dry ice and potassium iodide---table salt."

"TABLE SALT??"

"Yes, sir. This is what they use to seed clouds to create rain storms---not prevent a nuclear EMP attack!"

"Thank you, doctor. I've got what I need. Please put Colonel Blade on the phone."

Colonel Blade grabs the phone, "Yes, Admiral?"

"Colonel, I want Altus locked down. No one in and no one out. Make sure all employees and any personnel responsible for filling the tankers with the Chemtrail mixture are held for questioning."

"Yes, sir. We'll do it." They men hang up.

Admiral Whistler turns to General Jacobs, who is on the video screen,

"General?"

"Yes, Admiral. What do we know?"

Admiral Whistler replies, "This information is still being flushed out, but I have just received confirmation the tankers in Altus were not carrying the proper aerosol mixtures."

General Jacobs looks bewildered, "What do you mean? They were all carrying the Metrail mix. Aluminum oxide and Barium, among other things."

"No, General, they were not. They were loaded with a non-metallic mixture of potassium-iodide and dry ice. A rain cloud mixture. THE WHOLE DAMN COUNTRY IS EXPOSED TO ANOTHER EMP ATTACK!"

General Charles Jacobs responds, "This is impossible, I can't believe it! How did you find this out, so quickly?"

The General is now very concerned with the speed with which his scheme is being unveiled, piece by piece.

The Admiral says, "Teams of scientists and chemists are boots-on-the-ground, as we speak."

The General is surprised. "That was fast. How'd you get them there?"

The Admiral says flatly. "PAX River, Patuxent River Naval Air Station. We flew F/A-18F's up there and landed them right on Military Drive in Ft. Detrick. The scientists were on the move nearly two hours ago."

General Charles Jacobs is shocked with the speed of action and decision-making by the Admiral. "Well, this can only mean we were betrayed from within!"

"Yes, it does, General. Yes, it does. Don't you worry…we'll get to the bottom of this. And fast. In the meantime, I have ordered Altus, Marana and every other tanker capable to re-spray the sky with the metallic mixture. This time, the mix will be right. I don't need to tell you what would happen if any other weapons go off."

General Jacobs answers, "Yes, sir, understood. We'll provide full support to the Angel Shield fleet, of course!"

At 03:15PM Eastern Time, 10:15PM Qatar Time, Lori Sanders, Kristi Kruger and Brian Pierce are intently watching the tankers, on their playback monitors. They have slowed down the video feed, just to make sure they didn't miss anything in the infrared mode. Objects appear quite differently under this spectrum of light. They had to take their time to make sure they were doing their jobs correctly. The computer also flags anomalies, which have to be followed up."

Kristi says, "I see an eerie image of a seagull, lit up by the infrared. The El Toro and La Senorita must be near land." She wasn't paying attention to the fact the two vessels were nearing Cuba and Haiti.

Brian knows exactly where the ships are located. "They're just going through the Greater Antilles Ridge---they're only about twenty miles from Cuba. The water is pretty deep there it's over 10,000 feet in some places." Kristi says, "The birds look strange all lit up in soft, ghostly-yellow, just flying above the deck."

Lori Sanders startles them, "What's that? Look!" Lori points to a bright red spot off the port side of the El Toro. "It looks like a small engine moving through the water in the middle of the night! What's the time and tape say?"

Kristi reads back the date and time:

"October 6---02:21AM! Five days ago!"

Lori Sanders is getting excited, "Brian, bring up the other feed. What's happening with the La Senorita?"

Brian Pierce looks at the vessel, "Holy shit! It's there too! The same bright red spot is speeding to the vessel. It is also coming from the port side!" Brian continues, "Lori, nothing appeared on the deck when we reviewed the regular video tape. I mean, we checked that tape thoroughly, as did the computer. Do you think these guys are wearing active camouflage? You know, what our friends call, 'Sealskins?'"

Lori Sanders then says, "If they are, they won't show up on infrared, now, will they? But, if they *are* wearing Sealskins, they have special ultraviolet markers built into them---the exact frequency is top secret. Of course, DARPA and the NSA can track the frequencies from our satellites and this is also how the SEALs can track each other when they are cloaked. Switch over to the frequency on the UV spectrum. If someone, *somehow,* got a hold of Sealskins, they will be invisible on infrared, but they should glow like a peacock under this special, black light spectrum."

Brian switches to this spectrum and Kristi Kruger immediately says, "TWO BOARDING PARTIES! You can make out the figures of four men on each craft. Weird, they do look just like they are under a black light!"

Lori looks carefully at the water crafts, "They look like Zodiacs! Hmmm. OK, let's watch them board. We need to identify them." She continues, "Now, focus, let's see if we can use the facial recognition software to identify them. Look, they just sent a harpoon rope across the top of the deck. Here they come, climbing up. We should get good face shots of them, at least in the UV spectrum. See what you can zoom in on, Brian."

Brian rubs his hands together, "Yes, ma'am, here we go!"

Kristi looks at the images of the men, "Lori, they're wearing headgear---it looks like the Q-Warrior head gear issued to the SEALs! It might be difficult to identify them through their glass and helmets. Here comes the first one. *What* is *that*?"

Lori says, "What do you mean?"

Kristi replies, "I mean, look, carefully, over his right eye. Now look, he's looking *DIRECTLY UP* at the satellite---it's like he wants us to see him!"

Lori Sanders says, "What did you want me to see?"

Kristi points, "Look, he's wearing a pirate's eye patch! Now he's turning his head down, away from the satellite's eye. He's taking the patch off and throwing it overboard. Whoever it is, he has a sense of humor, I'll tell you that much. Lori, do you think it is Colt Jackson?"

Lori Sanders exhales deeply. "Let's not speculate, Kristi, let's try and confirm their identities...any of them."

Brian starts laughing his butt off. "These guys are great. Look! They're all wearing pirate eye patches! Same drill---they look up give us a face shot with their helmets and eye patches preventing their identities from being tagged, they then lower their heads and pitch the eye patches overboard.

"These guys are great!" Brian is crying his eyes out, he thinks it is hilarious.

Lori is trying to regain control, "Enough, Brian! Stick with it."

Kristi just murmurs, "This guy's a dream, Lori. I don't care what the Admiral says. Did you see that look of fierce determination and defiance on his face, today? Faced with the brig and full prosecution, he just spit in their eyes. What a man!" She exhales with a womanly pleasure of desire.

Lori Sanders can't handle it, "Both of you, BACK TO WORK! We will need to brief General Maxwell and Admiral Whistler on this ASAP. Finish the job and track their every footstep!"

The NSA analysts and their boss look at every inch of the footage from both vessels.

Brian Pierce recaps for all of them, "WOW! In, and out of both ships in 24 minutes. Looks like each team took out a patrolling guard and pitched the body overboard. These guys are amazing!"

Kristi thinks out loud, "I think the next time I see him, I'm going to ask him out on a date! He's so hot, he's smoking!!" Everyone can hear her.

Brian's great day, just turned into a disaster. The look of defeat is on his face.

Lori Sanders just rolls her eyes at both of them and starts to walk out to catch her breath. She thinks, "*This man, Colt Jackson, is really something.*" Before she leaves, she calls out over her shoulder, "You two, where are the Zodiacs going? Did they go to Gitmo? Can we follow them? I'll be back in 30 seconds. Freeze frame. I'm going to make a quick call."

Lori Sanders exits the pod area, but doesn't make a call. She is about ready to faint. She was wrong about everything. So was the Admiral and so was General Shane Maxwell---and so was the President of the United States!

The only person, who got it right, was Commander Colt Jackson. The man she loved, but treated like a school boy when she scolded and humiliated him before his superior officers. She wishes sometimes her quick mind would relax, a bit, and not react so quickly. This time, it betrayed her. She severely damaged a loving relationship and hurt a man she cared for deeply. She was certain this is also why he disobeyed direct orders from the Admiral.

She entirely pissed him off---as did the Admiral, as did the General and as did the President of the United States.

Fortunately, for the country, his actions saved the nation, or at least two-thirds of it. "*If we had only listened to Colt*," she thinks, "*not one nuke would have gone off!*" Lori returns to the pod area.

Brian tells Lori, "We can broaden out this feed to watch them leave, but we had the satellite trained on the tankers. Once they get a few miles away, they'll be out of range."

Kristi excitedly says, "Let's watch and see what happens! These guys are a dream!"

Lori sternly says, "Just do your job, Kristi. Please, no more cheerleading from the sidelines."

Kristi Kruger looks up, hurt from Lori's remark. "Sorry, Lori."

Lori responds, "I'm sorry, too. I didn't mean that. You're both doing a great job. And you know what? You're right---WHOEVER that is leading those men---HE IS A DREAM!"

Kristi takes a deep sigh. "He sure is, Lori, he sure is…WHOEVER he is. *HMMMM…I wonder*?"---she says sarcastically.

Brian has Kristi on the brain and just speaks out loud to himself. Both women hear him. "They're just studs. How can a guy like me compete with a guy like that, anyway?"

Kristi realizes she has hurt Brian. "Oh, Brian---you're the best! You're my Clark Kent, where would we be without you?"

Brian perks up, "Really? Gosh, thanks, Kristi." He smiles sheepishly.

Lori decides to give one of her sharpest analysts a vote of confidence on the woman front, too, "Yes, Brian. You're special and you have real talent. You'll find the right girl someday. I'm sure of it!"

Brian says, "Sorry about that, Lori. I was just thinking out loud. I do that too often, I think."

"No problem, Brian. We're all family here. It looks like the Zodiacs have slowed down about a mile away and they have brought the two craft together."

Kristi spots it first, "THERE! A helicopter!!"

Brian sees it hovering over the Caribbean, a thousand miles from nowhere and says, "HOLY SHIT! That looks like the new stealth S-97 Raider. *HOLY SHIT!* How cool is that?"

Lori warns, "Brian, enough with the swearing. We get it. This is quite an operation. All of it was done outside of the permission, or purview, of the United States Government. Maybe you're right. *HOLY SHIT!*"

Kristi adds, "I don't think I've ever heard you swear before, Lori! HOLY SHIT!"

All three of them laugh.

Lori gets back to business. "OK, let's grab this footage. Set it up and I'll send it over to the Admiral and General. Show it all, right to the point where they are picked up by the chopper, when the last man up scuttles the zodiacs. It looks like he had a knife in his hands. See if you can identify it before we send the footage to the Pentagon."

Brian says, "Yes, ma'am. I can zoom right in. The ambient lighting from the cockpit and lights from the chopper are reflecting nicely off it. Let me put the image through the computer and see what it comes back with."

Brian digitally grabs an image of the knife blade, just before it plunges into the first Zodiac. He then sends the image of the blade, the attached, black-gloved hand, and the top of the knife's handle into the NSA image database system for Weapons And Ammunition. Brian helped design the database and nearly every weapon on earth is profiled in it. He nicked named it WAM. Today everybody just calls it the WAM database.

Brian spits out the results, “Here we go. Looks like a 93% match.”

Lori Sanders has the answer before Brian can reveal it. “Don't tell me. It's a Microtech DOC Flipper Knife, right?”

Brian Pierce is shocked. “Just how in the *hell* did you know that?”

Lori answers, playfully, with a smile on her face, “Just a lucky guess.”

Kristi Kruger just stares in amazement at Lori Sanders. Lori leaves the pod area. She needs some time to think, before laying this all out for the Admiral and General. This man, Colt Jackson, will be the end of her. This she knows for sure---one way, or another.

At 04:30PM Eastern Time, October 11th, Lori dials the video conference for General Maxwell, who is still at the Pentagon with Admiral Whistler. She gives them the details of what they discovered from the infrared and UV video feeds from the Keyhole satellite.

General Shane Maxwell speaks, “This is going to go down as one of the worst days in American history, so forgive me for laughing but, PIRATE PATCHES! These guys do have a sense of humor---you've got to give them that.”

Admiral Whistler is dead serious, “Ms. Sanders. Were you able to positively identify any of them?”

“No, sir. We ran facial recognition and gating software, but with their suits and body armor on, there is nothing we could hang our hats on in courts-martial proceedings.”

The Admiral presses, “What about the Raider? Certainly we can get a trail of bread crumbs out of that?”

Lori answers, “There was one Raider approved to go on a drill, called ‘Deep Ocean Danger.’ We cannot identify if, whether, or not, this is the same Stealth Raider which picked up the people off of the Zodiacs. The other problem, sir, is with the Raider itself.”

What's that, Ms. Sanders? The Admiral asks.

“Well, sir, the Raider only has a range of 372 miles, without auxiliary fuel tanks. This chopper was OVER a thousand miles, round trip, from the U.S. Homeland. How in the world did they resupply it with fuel?"

General Maxwell says, "Check the airports in the Bahamas. Check Gitmo!"

"We've checked every single Navy facility and local airport in the Caribbean. There is no record of a helicopter, of any kind, refueling during the evening of October 5th into the morning hours of October 6th off of the U.S. mainland. The Raider participating in the 'Deep Ocean Danger' training drill last refueled at the Ft. Lauderdale Naval Air Station on the 5th. After that, it apparently just hopped back up the coast to Virginia Beach the next morning, having completed its training mission."

General Maxwell requests, "Ms. Sanders, get us the base footage of the landing of that chopper. I want to know who the pilot was and who else was on board."

"We did that, already, sir, anticipating your request."

"And what did you find out?"

"The helicopter was being flown by a distinguished Afghanistan war veteran by the name of Commander James Forrestal. Admiral, he's actually the brother of one of your "new kids on the block" crack technologists."

Admiral Steven Whistler is surprised by this coincidence, "You mean, Lieutenant Ronnie Forrestal?"

"Yes, sir. Ronnie worked with Colt Jackson and our team to help analyze the contents of Evan Barrister's encrypted thumb drives---remember, Evan labeled them, 'The Father, 'The Son' and 'The Holy Ghost,' during the attack on the financial system. He, Lieutenant Izzie Brown and Lieutenant Jake Sampson did great work."

Admiral Steven Whistler buzzes his intercom to his adjutant. "Bring Lieutenant Ronnie Forrestal up here, right away. And order Commander James Forrestal to also appear before me, immediately. Have him fly up ASAP."

Lori Sanders adds, "Sir, there is one more thing."

"What's that, Ms. Sanders?"

"The last man out of the water scuttled the Zodiacs with a DOC Flipper Knife. While we didn't get a clear shot of any of their faces, we nailed the knife identification through the WAM database. It's a 93% match. I think the hand enclosed around the handle prevented the match being 100%."

The Admiral exclaims, "A Microtech DOC Flipper Knife? You mean the same knife Colt Jackson used when he single-handedly took down the Russian mercenaries inside the Federal Reserve last year, during the Blood Moon of the D'Apocalypse?

"Yes, sir. The same one indeed."

The Admiral concludes, "Well then, that should be enough evidence to nail his butt to the wall, shouldn't it?"

Lori states, "Well sir, I'm not quite so sure about that."

"And why is *that, MS. SANDERS*?" The Admiral is growing frustrated with Colt's clever ability to cover his tracks, with style, humor and brash bravado.

Lori continues, "After the incident at the Federal Reserve, out of respect for Commander Jackson, every member of every single SEAL Team, put in for an exchange of knives. Many were denied for budgetary reasons, but many weren't. Furthermore, it was learned through our credit card databases for military personnel, many hundreds of U.S. Navy SEALs purchased the knives with their own money, when the government didn't approve the requests."

General Shane Maxwell speaks up, "I hate to quote General George S. Patton at a time like this, but, 'What a magnificent bastard!'"

Admiral Steven Whistler says, "If I didn't love him so much, I think I'd personally shoot him. I still can't believe he directly disobeyed me, a personal direct order. If we let the military be run like this, we would have chaos."

The General looks at the Admiral, "Obviously, this situation is delicate. What do you think about briefing the President? I'd like your opinion, also, Ms. Sanders---you'll do the briefing if the Admiral concurs."

Admiral Whistler says, "I do agree. I'm simply not objective about this matter. I'm still spitting nails over it. Right now, we're going to leave Commander Jackson in the brig to let him think about this. I'll call the President"

The General says, "Ms. Sanders, wait right there on the line. The President will likely want to take this briefing, immediately. Be precise and quick---the man has a lot of pressure on him, at the moment."

"Yes sir, I will."

At 04:55PM Eastern Time, Admiral Whistler makes the call to the President. The President is immediately bridged in. Lori Sanders has explained everything to President John Diaz. The President wants every detail, he listens intently.

President John Diaz says, "Yes, Ms. Sanders, go on."

Lori continues, "And then, sir, at the very end we saw the knife which was made famous by Commander Jackson at the Fed. It was the same one he used to take out the Russian Merc, when the Commander was near death. Do you remember?"

"How can I forget it? What did the papers say, 'DOC Flipper Knife, For Life!?' It was something ridiculous like that when word slipped out about details of the heroics displayed by Commander Jackson."

Lori apologizes to the President, "Sorry sir, I know you were there also. I just thought you might have a lot on your mind at the moment."

President Diaz replies, "Not at all, Ms. Sanders. In fact, I appreciate the details of your briefing. So, Admiral, can I net this out?"

"Of course, sir."

The President sums it up, "Commander Colt Jackson sat in our intelligence briefings, some of which I attended also, objected vehemently to our collective strategy and became a rogue force against direct orders. Furthermore, it appears he either stole, or commandeered some of the most sensitive and highly classified United States military equipment in our arsenal and commanded military personnel to implement a plan which was completely off the reservation, compromising national security protocols, across the board. It's pretty tough to run an army this way---is that about it?"

Admiral Whistler can't agree more, "Yes, sir."

The President asks the General, "Do you agree, General?

"Yes, sir."

Lori interjects, "But sir, Commander Jackson was the *ONLY* one who was right. Do we even know what happened to L.A.---why didn't Angel Shield work?"

The President looks at Ms. Sanders, "That is an excellent question, Ms. Sanders. Admiral, what do we know---anything yet?"

The Admiral answers, "The General and I were going to brief you, just after this meeting."

The President states, "I think we should all hear the news, if there is any. Fire away."

Admiral Whistler explains what he believes has transpired, "Sir, it appears the United States of America has suffered an act of sabotage."

"Admiral, this is no time to cover your asses by claiming sabotage if Angel Shield doesn't work. But, for the moment, I'll give you the benefit of the doubt. Please proceed."

The Admiral looks over to General Maxwell, somewhat chagrinned by the President's remark. "Immediately after the attack, we flew teams of scientists and chemists out to Altus and Tucson. They confirmed nearly all the planes flying over the United States in the Angel Shield tanker squadrons on the morning of October 11th were NOT carrying the correct mixture of Chemtrail materials to create the metallic mesh for Angel Shield."

The President queries, "What were they carrying?"

"They were carrying potassium-oxide and dry ice. These are ingredients used to make rain. They also create clouds. Only Northern California had the correct mixture of Metrails in its tanks. This is why Silicon Valley, San

Francisco, Sacramento and the area basically above Santa Barbara were unaffected by the attack."

The President asks, "We're SURE of this?"

General Maxwell states, "100% positive, sir. We have also discovered employees of a third-party supplier of the materials, are missing. There are four of them, to be exact. Two are employed at Altus and two are employed at the Marana Air Park in Tucson. They are Chinese-Americans, sir. We're looking into it, but there appear to be a large number of phone calls to Beijing in the days leading up to the attacks."

The President is incredulous, "You've got to be kidding me? Seriously?"

The General answers, "I kid you not. I would suggest a call to the Premier would be in order sir."

The President jumps on this, "You bet your ass it is. I know the Joint Chiefs have got us at DEFCON 1---keep us there. Keep our pistols cocked. We really don't know everything, at all, yet. Let's see if this Chinese connection is the real deal. Maybe this is a whole lot bigger than the Islamic State."

General Shane Maxwell answers, "Yes, sir."

President John Diaz is thinking for a moment, looking off in deep thought, while on the video display.

The Admiral prods the President, "Mr. President? Is there anything else?"

The President answers, "Yes. We're all damned fools. We have this evidence of sabotage and we're holding the only person on earth who warned us about our assumptions in the brig! He directly *warned us* that our assumptions were based on perfect test conditions. YOU DO REMEMBER THIS, *DON'T ALL OF YOU*?"

Lori, the Admiral and the General immediately respond, "Yes, sir."

The President goes on, "So, you've locked up the ONLY person who probably prevented a complete wipe out of the entire United States' infrastructure. You know we don't have enough transformers for the substations, as it is. If we hadn't, at his urging, torpedoed and destroyed the three cargo ships out of Amsterdam, and if those two other tankers hadn't been intercepted by Commander Jackson---whether he admits it, or not, it would have been the doom of our entire nation. What kind of crap is it to keep a man like this in jail! If I could give him another medal, I would!"

The Admiral asks, "What are your orders, sir. He directly disobeyed me and General Maxwell, created a rogue operation with unknown co-conspirators..."

The President thunders, "AND SAVED THE NATION! RELEASE HIM, IMMEDIATELY, ADMIRAL, OR I WILL RELIEVE YOU OF COMMAND! Besides that, Admiral, you don't have enough proof. That knife is the most popular knife in the SEALs today. Unless you can find a matching pirate's eye patch in Colt Jackson's Georgetown condominium dresser drawer, get him out of the brig, now!"

Lori Sanders stifles a laugh.

Admiral Steven Whistler immediately replies, "Sir, yes sir!"

The President then calms down a bit, "Gentlemen and Ms. Sanders, we are all under a great deal of stress. Why don't you bring the Commander over to the White House and all of us will plan what, exactly, we're going to do to get these guys in the Islamic State. We need to pulverize them and get every one of the sons of bitches who are responsible. We know their key leadership, correct?"

Admiral Steven Whistler replies, "Yes, sir."

The President concludes, "Excellent. I'll see ALL of you here in one hour. Does that give you enough time, Ms. Sanders?"

General Shane Maxwell tells Lori, "Ms. Sanders, have the helicopter take you to the Pentagon. We'll prepare for the President and go over from here."

"Yes, sir." They end the call.

At 06:00PM Eastern Time on the 11th of October, the team gathers in the Oval Office at the White House. Lori Sanders, Admiral Steven Whistler, General Shane Maxwell and Commander Colt Jackson are meeting with the President.

After five grueling hours, the plans have been drawn up. The group has been working, non-stop. Colt Jackson is a pure professional. He doesn't let his trip to the brig, or the cross words thrown at him by the Admiral, affect him.

Colt Jackson tells the President, "The Emir and the Caliph are the key, sir. From these two sources, the Islamic State has grown in power and influence throughout the Levant. They now control most of the Eastern Mediterranean."

Lori Sanders says, "I agree. The analysis of all the interactions at the Marsa Malaz Hotel, among the attendees, clearly indicates these two are in charge."

Admiral Whistler summarizes a key component of the plan, "Sir, the 5th Fleet is right in Bahrain. The aircraft carrier USS George H. W. Bush (CVN 77) has plenty of fire power. We can blast the Islamic State to kingdom come. Because of their use of nuclear weapons, sir, we are allowed in all fairness, to use nuclear weapons to strike back. We can even use neutron weapons, which won't leave radiation behind. We can really just kill the enemy."

General Shane Maxwell concurs, "I agree sir, it's about time we took the gloves off and went after these bastards. They are like rats. If they are not attacked, they will multiply and continue to raise havoc, worldwide."

Lori Sanders also agrees, "Sir, I believe the people of this country are dying for payback. No one likes murdering innocent people, but in an age when nuclear weapons can be used in a sneak attack, as was amply demonstrated today, we must strike back. It wouldn't be too much to even strike back a hundred times more aggressively---just to show these blood-thirsty people what the consequences will be if they continue their murderous ways. By last count, sir, at least one hundred aircraft crashed today, as a direct result of the EMP. Thousands died."

The President turns to Colt Jackson, "Commander? You have been more quiet than usual this evening, which is perfectly understandable considering the debt of gratitude this nation owes you---and the way you were treated. But, I am asking you to put this behind you. We have to go forward as a nation."

Colt Jackson looks the President in the eye, "Sir, I am fine. Admiral Whistler was just doing his job. Although, I admit to being really ticked off at how I was treated by him, General Maxwell…" Colt then looks directly at Lori Sanders, "…AND Ms. Sanders. However, I will always put this country first. All is forgiven. I think it was all a misunderstanding. By the way, did you ever positively identify anyone who *may* have sabotaged and intercepted the tankers?"

Lori clears her throat, looks Commander Colt Jackson in the eye and answers, "Who said anything about eight men, in the middle of the night, on two different Zodiacs, with over 10,000 feet of water beneath them, climbing aboard speeding oil tankers, wearing pirate eye patches, while flipping us "the bird," and looking up directly at the satellite when each of them clambered on board. These nameless men then sabotaged the ignitions, and wiring, on two separate SA-2 missiles, using identically shaped charges. The charges just *happened* to be of the perfect size and precision to NOT blow up either the liquid-fuel tank, or the solid-fuel rocket booster, while causing *NO* damage to the 10-kiloton nuclear weapon, sitting on top of the rocket. Finally, these mystery men, each took out a roving security guard without alarming anyone on the entire ship. Then, they scaled back down the harpoon rope, got on their Zodiacs, rendezvoused with an S-97 Stealth Raider helicopter---right off the coast of Cuba, once again in over 10,000 feet of water. Only after every other man was plucked out of the sea by the Raider, the last man in the water pulls out a Microtech DOC Flipper Knife and plunges it into the Zodiacs to scuttle them---presumably just before he was going to be eaten by a *SHARK*!"

Lori pauses for effect. "*Hmmm…let me think…*did we ever identify the men responsible for undertaking this mission, which took a total of only 24 minutes?" Lori again pauses, shakes her head, and says, "Unfortunately, no Commander, we could not prove, with certainty, their identities."

Colt Jackson rubs the growth of his beard, and then he replies, sarcastically, "*WOW*, Ms. Sanders, *24 minutes…*Those guys *must* be pretty good! What a *great* analysis. *But I was wondering* why you think they flipped you 'the bird?' It sounds kind of personal, to me. Too bad you have to see and analyze these kinds of operations, after the fact. If you want me too, I can ask around, maybe some of the guys know who was responsible. I can let you know if I learn anything."

Colt smiles and cocks his head.

Lori Sanders stares at Colt. She rolls her eyes and says, "I won't hold my breath, Commander."

Admiral Steven Whistler says, "Knock it off, Colt."

General Maxwell says to Lori, "You too, Ms. Sanders. Let's get back to work."

Colt looks at all of them, "Sorry gentlemen. Excuse me Ms. Sanders. And I'm sorry, Mr. President. I had to get that off my chest."

Lori Sanders asks, "Are you about done, now?"

Colt laughs. "Yes, ma'am, I sure am." He turns to the President. "Sir, I have an idea, it's a little different. I also don't know how it will be received, because it's pretty aggressive."

President John Diaz has seen that look in Colt Jackson's eyes before. He asks, "More aggressive than using neutron weapons?"

Commander Jackson replies, "In one sense, yes sir."

Admiral Whistler apologizes to Colt, "Commander, IF you did undertake that mission on the tankers, I'm sorry for the way I treated you, today. You did a brave thing, filled with valor. I'm really sorry, Colt. If you didn't do it, then I also owe you an apology."

Colt looks at his friend and superior officer, "No problem, sir, either way."

The Admiral asks Colt, "What's your idea, Colt?"

"Well sir, you actually gave me the idea by mentioning the 5th Fleet. Between those forces and the forces we have in the area, including Kuwait, Saudi Arabia, and even Yemen, as limited as they are everyone understands we have a formidable presence.

The President probes, "What do you suggest?"

Colt answers, "Sir, we have always prided ourselves in delivering a fair and appropriate response when we have been attacked---whether it was at Pearl Harbor, or during 9/11. But, in this case, it was an EMP attack. Ms. Sanders is correct. Thousands of people lost their lives, today, because of this nuclear device, but not because of a massive shock wave. It was caused because the electronics were fried. In fact, all the electronics, for practical purposes, are fried throughout Southern California.

President Diaz has some news on this front, "The good news is we already have thirty replacement substation transformers heading west. We had twenty stock piled in the limestone caverns of Missouri and we are going to have every electric utility company chip in to replace the rest. Southern California will have power within a couple of days."

General Shane Maxwell is surprised and elated, "Congratulations, Mr. President. This is wonderful news."

Colt continues, "This brings me back to my point. The impact from the EMP detonation is primarily economic, notwithstanding the thousands of poor souls we lost today. It seems to me the best way to get even is to take out those responsible. Send in Special Forces to take out the radical leaders---by whatever means necessary. This could be by sniper, poison darts, or direct confrontation. Every time one of them pops there head up, we will be there to take them out. It will take some time, but I have found the fire in the belly of a radical Muslim quickly goes out when he sees his leaders quickly die. They get the message very soon and crawl back into their holes, like the rats they are."

President Diaz asks, somewhat surprised at Colt, "That's it? That's the plan? I don't know if that will be enough, Colt."

Colt, again continues, "Mr. President, excuse me, sir, I didn't quite finish."

"What else?" asks the President.

"We need to take over Qatar and force them to repay double the amount of the losses we will experience in Southern California. The top ten companies in the region, alone, had revenues of nearly $250 Billion last year. They will take months to recover. I'm sure the NSA and CIA can come up with the real numbers, but I'm betting, overall, the damage will be close to $1 Trillion. This gives us the perfect reason to take over this country."

General Maxwell looks to the President and states, "Mr. President, the Commander makes an interesting point."

"Sir, their Emir has been caught smack, dab in the middle of it." Colt goes on, "...so has his Admiral and so have the tankers from LUBEROL---as ordered by QEMCO. Each was loaded with yet another 10-kiloton weapon. Navy divers, I'm certain, can recover the evidence. But, right now, sir, there isn't a government on earth which will stand in our way. We need to blitzkrieg Qatar and take over---just like we did in the Philippines, and just like we did in Japan, and just like we did in Germany, at the end of those wars."

The President looks at the Admiral, "Admiral, what do you think?"

"It will send a shock wave through the Muslim community. While the radicals have sabre-rattled their way to the Islamic State, this takes away most of their money. It also provides a legitimate stronghold for our economic and social interests in the Middle East. Commander Jackson is correct, if we are going to do this, we should do it immediately."

The President decides, "Gentlemen, I want a full-scale attack implemented immediately, as soon as you can make it happen. Commander Jackson, if you would like the honor of going after the Emir yourself and arresting him, to be arraigned here in America on murder and conspiracy charges against humanity, he's all yours."

Colt Jackson pauses, looks at the President and replies, "No sir, I cannot."

Lori Sanders, still in love with Colt Jackson, looks up in surprise and in relief. She thinks, "*Finally, her man is not going into battle*!"

The President is surprised, also. "I'm surprised, Commander, but if any man deserves the right to stay out of this fight, it's you. You've done enough."

Colt goes on, "No sir, I think you misunderstood me, sir. I mean I won't do it alone---but I have *seven* other guys who would follow me to hell and back and I wouldn't dare go on this mission without them." Colt turns his head from the President and stares directly into Lori's eyes.

Lori Sanders returns his gaze and says, "You are something else, *mister*."

The President appraises Lori Sanders' remark and then he laughs, really hard. "Yes, Colt Jackson, you really are something else, and this country is honored to have you serve it."

Admiral Whistler concurs, "Amen to that, Mr. President."

General Shane Maxwell says, "Touche, Mr. President. Please thank your men, Commander. The country owes them a debt of gratitude, also."

Colt continues to play the game, "For what? They haven't done anything yet."

The President orders Colt, "Go home and get some sleep, Commander. You'll have to tell me about the Raider someday. It might make a good replacement for Marine One and I'd like to hear what you have to say about it."

Colt replies, "Anytime Mr. President. And sir, thank you for standing by my side. I'll never forget it."

The President's dark brown eyes fix themselves on Colt. He simply says, with humble sincerity, "Thank you for standing by ours, Colt."

At 10:50PM Eastern Time, the meeting breaks up and Colt grabs an apple from the fruit bowl in the Oval Office, he didn't think the President would mind.

They walk together out of the Oval Office, past the Rose Garden and head out the West Wing. Colt has gone ahead of the Admiral, General and Lori Sanders, and exits out the White House gate onto West Executive Boulevard. Colt walks toward Pennsylvania Avenue, where he'll make it to 17th Street NW and grab a cab home to Georgetown. Lori excuses herself from the General and Admiral and says goodnight. She rushes to catch Colt before he disappears into the night. She grabs him by the arm and Colt spins around.

Colt Jackson is surprised to feel someone grab him, "Lori?"

Lori pleads, "Colt, PLEASE forgive me. I'm so sorry."

Colt replies, "I already forgave you the moment it happened. Don't worry about it. Everything will be OK."

Lori continues on, "Have you had time to think? Can we just go back to Denver? *Please*, Colt."

"I have a lot on my mind, Lori. Even you have to admit, it's been a pretty rough day."

"I know. I'm sorry Colt. Can I ask you just one question---it's been bugging me the whole time."

Colt Jackson stares deeply into Lori's sparkling green eyes. She looks beautiful in the evening light.

Colt sighs and answers, "Of course, you can. Go ahead."

Lori melts in front of him and wants nothing else but to kiss him, again. She longs for him, but has to ask him just this one question and stutters a bit---she's nervous, "Colt, I..uh...uh...you know how stupidly anal I am, and my brain gets me into more trouble than most, but something is really bugging me. I mean, I figured the whole thing out. I even know about the connection to Prince Harry with Commander Forrestal, which I might add, I left out of the report."

Colt replies, "Thanks for that, I guess. What's your question?"

Lori asks, "How did you refuel the Raider? It's impossible to fly from Miami to the far end of Cuba, drop off a SEAL team, then hover around for a good thirty minutes, pick up a SEAL team, and then head ALL THE WAY BACK to the States! How did you do it?"

Colt feels his anger well up, again, "That's what you wanted to ask me? After everything that's happened?"

Colt is getting visibly upset, again, with her. He pauses, trying to regain his composure, but fires back a snide response. "You know, with your talent, drive and ambition---I'll bet if you study the Keyhole video feeds long enough, from every single satellite around the world, you'll get your answer---and the promotion you've been aiming for. But until then, I can't confirm, or deny, anything, ma'am. YOU, HAVE A GOOD EVENING. GOOD NIGHT!"

He pulls the red apple up to his mouth, takes a big bite, turns around, shakes his head, and storms off into the evening.

Lori cries out to his back, "Colt, PLEASE! Don't leave. I didn't mean it like that! Please!"

Colt leaves briskly at a half-trot to 17th Street. He hops into a cab.

He's gone. Lori stands about twenty yards behind him, sobbing. She has lost her man. She hopes and prays it is not forever.

CHAPTER 22

THE HAREM & THE PALACE GATE

At the Emir's Palace Gate of Amiri Diwan Al Bidda in D'oha, Qatar, it is October 25th at 03:00AM Qatar Time and 8:00PM Eastern Time on October 24th.

Commander Colt Jackson is leading a contingent of men to take over the palace of the Emir of Qatar. Colt has taken charge of SEAL Team 3, which covers the Middle East and is specially trained for fighting in desert climates. The Team has one-hundred-twenty men, and is broken down into three forty-man Troops. Colt has brought along his "Heaven's Seven" with him for the mission.

Of course, these are the men who would follow him into hell and back:

> Lieutenant Cody Custer (call sign "Tiger Fox");
> Chief Ryan Duffy (call sign "White Shark");
> Chief Spider Sinclair (call sign, "Black Widow");
> Master Chief Bull Sherman (call sign, "Sierra Hawk");
> Ensign Dizzy Lee (call sign, "Stone Fish");
> Master Chief Jack Flintlock (call sign, Killer Whale");
> and Petty Officer Duke Moran ("call sign "Stingray Charlie").

They are also men who have become legends. Rumors are quickly passed along among military people, and word spread like wild fire that it was Colt Jackson and his men who were responsible for taking out the missiles on the *cargo* ships on the East Coast which would have destroyed America. To this day, their involvement with taking out the tankers from Venezuela has been kept a complete secret from the public. This bit of sleight of hand and spinning of the story protected two things. It protected the identity of Croc Hunter---General Jose Lopez, the secret double-agent working for America in Caracas, and it also kept the involvement in the operation of Ghostship completely top secret. The USS Minnesota nuclear attack submarine is one of America's prized defense assets. The United States never confirms, or denies, anything she does.

The military newspaper, "Stars and Stripes" even ran an article about the rumored heroics of an unnamed SEAL Commander and his "Heaven's Seven" brigade of righteous warriors. Its Op-Ed page claimed they wielded the sword of Gabriel to save the day, by secretly attaching mines to the cargo ships, and destroying three nuclear missiles in the process. After all the press and the major spin on the story to protect the General and Ghostship, the name "Heaven's Seven" has stuck, ever since.

If any of the SEALs are ever asked about their involvement in the mission, they simply reply, with a smile on their face, "Sorry, ma'am/sir, but I can't comment on national security matters..." It drives the press crazy, but the guys always get a good laugh out of it.

Now, in Qatar, these men have taken key roles under Commander Jackson's command. When SEAL Team 3 heard the Heaven's Seven was joining them for this mission, they greeted them with wide-open arms to take down the Emir of Qatar and his entire government.

All the men are wearing land versions of Sealskins, which make them virtually, disappear in the visible light spectrum. Stealth is their trump card. Snipers are posted in the clock tower on the palace grounds, as well as along the walls of the Emir's living quarters, which are massive. The blue light bouncing off the white walls, in the evening, makes the palace a spectacular sight. Their mission is to abduct the Emir and pave the way for the U.S. Marines and components of the U.S. Army, to seize D'oha and all the oil and gas wells in the country. The Air Force and Navy will destroy the Qatari Air Force and Navy. Their operation will commence in in approximately two hours.

Each platoon leader has his video cam turned on and ready to begin the assault.

Colt Jackson speaks into his headphone. "Snipers---on my mark. 3...2...1 FIRE."

Three squads of snipers, each consisting of four men, are strategically positioned around the top of the Emir's palace walls, as well in the Clock Tower outside. To

Clock Tower and Amiri Diwan (Emir's Palace)

Al Bidda area. Doha, Qatar from OnlyImage.com – With permission from onlyimage.com. See more at: http://www.onlyimage.com/photo/09-clock-tower-and-amiri-diwan-emir-s-palace-al-2960159#sthash.gwfYJxZz.dpuf

eliminate the Qatari guard, which are both inside and outside the palace walls, the three squads of men have taken deadly aim with their FN MK20 sniper rifles---all equipped with silencers. They Fire. Twelve bodies collapse. The snipers continue sending rounds into standing soldiers until there is no one left standing. Every shot was taken in deadly silence.

There are no alarms going off in the Palace. The Emir is still sound asleep with his harem.

Sniper platoon leader Snake Eye hails Papa Bear over the COMM, "Papa Bear, this is Snake Eye, all is clear. We count twenty-five down. Over."

Colt Jackson replies, "Snake Eye, read you. We're going in, get prepared for the breach of the palace door. We've just cut the video surveillance, as well. We don't want dead bodies being broadcast out to the rest of the country---at least, not yet."

Sniper Platoon Leader Snake Eye responds, "Roger that, sir. I'll tell the men to get ready."

Chief Spider Sinclair and Master Chief Jack Flintlock sprint to the door and set the C4-shaped charge explosives in place. Over one hundred Navy SEALs will soon storm the palace. With their invisibility cloaks active, the raid and capture of the grounds will begin in approximately one minute.

Inside the White House Situation Room Lori Sanders, The President and his staff, Admiral Steven Whistler and General Shane Maxwell look on. The rest of the Joint Chiefs of Staff is also present around the table. General Charles Jacobs is plugged in from Colorado Springs, inside Cheyenne Mountain, at his NORTHCOM Command Center. They are all watching Colt Jackson and his Navy SEALs attack a sovereign nation with the purpose of conquering it and remaining in it, until the Arab Spring is crushed.

It will then become an Arab Winter.

General Maxwell asks Lori Sanders, "We're all set with jamming the palace's frequencies?"

"Yes, sir. They won't be able to make a single call."

Inside the Emir's palace, it is now 03:20AM Qatar Time, 08:20PM Eastern Time.

Black Widow---Chief Spider Sinclair---gets the honors. "Up yours, you miserable bastards!" The Chief crushes the detonator button with his thumb. Immediately, there is a huge explosion, and the doors of the palace entrance are blown inward by the C4 blast.

Chaos immediately reigns inside the palace, as more guards come out with their guns raised. They point them frantically, looking for the intruders responsible for the blast. They see nothing. The guards are immediately dropped in place, as the snipers continue to use their silenced FN MK20's to full advantage.

Colt has his men prepared for close combat. They also are using silencers and they begin the process of clearing the building and taking prisoners.

Colt Jackson turns to his men, "This is it. This is the Emir's master bedroom, according to the floor plans. Drug him, bind him and extract him. But, be alert. Many of his women are trained assassins. They may have weapons, but just as easily, they can kill you with a swift kick. Do not hesitate to pull the trigger, if you have to."

Colt's men reply seriously, "Yes, sir."

The men open the door. It is unlocked, because the Emir's guard is supposed to protect it, but they have all gone into the court yard, where they were easy prey for the snipers.

The SEALs get ready to enter, but before doing so, Colt orders them to reactivate their Sealskins. They slowly open the door and see the grotesque sight of a naked, fat Emir Sheikh Muhammad Al Therani covered by six naked women. The men of Heaven's Seven begin pulling the women off. One of the women attacks the invisible hand grabbing her body with a forceful leg kick.

Ensign Dizzy Lee, call sign Stone Fish, immediately responds, "You b-e-e-e-a-c-h!" He doesn't hesitate to fire, per orders. The woman drops, shot right through the forehead.

The rest of the women move away from the Emir and are corralled into a corner. The Emir is still sound-asleep, snoring away. A couple of bottles of 50-year old Macallan Scotch Whiskey are within arm's length of him. Colt recognizes the bottle as the same one Daniel Craig drank from in the movie, "Skyfall." They cost about $48,000 a bottle. One is completely empty and the other is half full. The man is completely passed out.

Back at the White House, the scene is watched through the SEALs' video cams, with fascination, as Colt Jackson and his men snatch the naked Emir away from the unclothed bodies of the naked women. The fat ruler literally had surrounded himself with his harem, while he slept. It gives Lori Sanders the creeps. She shudders.

Lori spits out, "What a complete pig!"

Admiral Steven Whistler shakes his head, "Absolutely disgusting."

Back at the NSA Glass House in Ft. Meade, Kristi Kruger and Brian Pierce were cleared by Lori Sanders to observe the SEALs in action during this mission. Kristi is repulsed.

Brian is fascinated by what they witness.

Kristi talks to the screen, "That is absolutely the sickest thing I've ever seen. What debauchery."

Brian is a bit flip, "It certainly is, but I must admit, what a lifestyle!"

Kristi tries to exhort him, "Brian!"

Brian says, "I'm just kidding around, Kristi! Geesh!"

At the Emir's Palace, it is 03:35AM Qatar Time, 08:35PM Eastern Time. Colt Jackson and his SEALs are fifteen minutes into their raid on the Emir.

Colt, observing the completely sinful surroundings of the fat Emir talks, off-handedly to his Team, "I wonder what Allah has to say about this, now? Nothing like the light of discovery to uncover the darkness."

Tiger Fox replies, "I don't know Papa Bear, but the hypocrisy is simply amazing. They criticize us for our ways and look at this!"

Stingray Charlie, Petty Officer Duke Moran complains, "Yeah, we can't even have a drink when we're stationed here and look at this guy. Drunk as a skunk, with James Bond whiskey, to boot!"

Commander Jackson orders his men, "OK, drug him, tie him up and we will get him on a UH-60 Black Hawk. I've already signaled them to come in. They'll pick him up and bring him back to face justice and they'll be here in three minutes. Let's go. Once he's in the air, we wait for the Marines to take over the compound and its command center. After that, my friends, we are off to grab the Caliph---which won't be quite as easy. YOU KNOW THE PLAN."

Colt's men give him their oral salute, "HOOYAH, SIR!"

Inside the White House Situation Room, a look of concern comes over Lori Sander's face. She looks at Admiral Steven Whistler and General Shane Maxwell, somewhat bewildered.

Lori Sanders asks the two men, "What plan?"

The President looks at the Admiral, "Yes, Admiral, what plan?"

Admiral Steven Whistler responds, "It's just an operational detail, sir. It's something I worked on with Colt during the last couple of weeks."

Lori Sanders presses on, not liking being left out of this part of the operation, asks somewhat urgently, "WHAT'S THE *PLAN*?"

CHAPTER 23

THE "PLAN"

At the White House, Admiral Steven Whistler begins to unveil the rest of the plan to take down the Caliph and the Islamic State. It is 08:40PM Eastern Time on October 24th and 03:40AM in Qatar, on the 25th of October.

The Admiral explains to Lori Sanders (who asked the direct question about what the "plan" is) and to the rest of those gathered inside the Situation Room, "Well, it's pretty simple. Colt figured it would be pretty tough to smoke out and then decisively eliminate a significant number of Islamic State Jihadists. They all hide among civilian populations, and he thought they would protect the Caliph at all costs, in a direct attack. A traditional ground war could be brutal. The door-to-door action, with complete confusion in trying to identify innocent civilians versus real bad guys, would inevitably cause heavy civilian casualties. City fighting, as we all know, also causes very high casualties for our troops. Bad guys go in and out of their rat holes, seemingly at will. It could get pretty bloody, on many fronts."

The President chips in, "He makes a pretty good point. What did he suggest, precision bombing?"

Admiral Steven Whistler looks solemnly at the President, "No, sir. You're familiar with General George Armstrong Customer and the Battle of the Little Big Horn?"

The President nods his head and says, "Yes, according to my history books, it didn't end so well for the 7th Calvary."

Admiral Whistler affirms the President's knowledge of history, "Right. They were wiped out by the Indians just up from the Little Big Horn River in Southeast Montana, about 100 miles from Billings. The attacks took place over a two-day period from June 25th-26th in 1876."

Lori Sanders is growing beyond impatient. She didn't care what the Admiral's rank was, "So…what's the plan?"

The Admiral answers, "Colt reasoned he's a lot like General Custer. In fact, Ms. Sanders, you'll be interested to know he referred to himself as being somewhat 'thick headed' at times."

Lori Sanders feels sadness roll over her, she says, "He did?" She looks embarrassed Colt would use her words to self-describe himself.

Admiral Whistler continues, "I prefer to think of him as determined. In all the years I've known him I have seen him use that thick skin of his to make incredibly brave---but never rash, decisions. Well, almost never---the Federal Reserve mercenaries aside."

The room has a moment of light laughter. Everyone will never forget Colt's heroics at the Federal Reserve in New York when he was vastly outnumbered by Russian mercs.

"Anyway, his plan is to take his one-hundred-twenty men, led by his Heaven's Seven, and draw out the Caliph."

The President raises his eyebrows in a frown, "Just how does he propose to do that, pray tell?"

Admiral Steven Whistler answers the President, "Before he went back to Qatar, he made a little stop at a place called Maki. You remember that place, don't you, Ms. Sanders? You and Colt had dinner there. Afterwards, your team performed the background investigation into Paul, the manager of the Maki, remember?"

General Shane Maxwell now joins the explanation, "Yes, Lori, Paul was clean, but then you started digging deeper and you realized the Islamic mole at Maki, may be a person no one would suspect. What was her name, again?"

Lori Sanders immediately responds, "Akiko. Yes, it was her. We left her on a string to monitor her. She can be picked up any time. She nearly had Colt murdered."

Admiral Steven Whistler is stifling a chuckle, "That's not what Paul told Vice Admiral John Stiller, call sign 'Blue Angel,' and Commander of the 5th Fleet in Bahrain. John told me the assassin did indeed try to kill Colt, but Colt clobbered the killer, IN THE BATHROOM, and then tied the would-be assassin to the toilet. Apparently, afterwards, Colt just went out to his table, cool as a cucumber, finished eating his meal---with you, Ms. Sanders, and then casually left."

Lori Sanders feels somewhat busted on this, because she never reported the incident in the bathroom. She offers, "I honestly had no idea what had happened, until I saw blood on him at the Moda Mall after dinner. We were just walking around. I made him tell me what happened. After the op at the Marsa Malaz, I dug pretty deep. I started going through all the cell phone calls made from Maki, while we were eating dinner. One, in particular, caught my attention. I remember it vividly. It was made right when we left. It was a woman's voice, with a Japanese accent---'Hello, Hassan? Muhammad is not answering his phone and Commander Jackson and his date just left the restaurant.'"

General Maxwell confirms what Lori discovered, "Exactly, Lori. Anyway, you're the one who figured out Akiko is an ISIL spy."

Lori responds, "I didn't think anyone did anything with that piece of information, yet."

Admiral Steven Whistler, looking at her somewhat sternly, says, "Well, you're 'thick-headed' Colt Jackson decided she was the perfect pawn in this chess game."

Lori Sanders replies, "He did? What did he come up with?"

The Admiral replies, "Well, he went back to Maki and had dinner with Admiral Stiller, a few days after our meeting with the President on October 11th---the day of the EMP attack. He, Colt and I have dinner together, every now and then. Believe it, or not, we're all friends and go to church together, sometimes."

The President is also growing tired of the background. He wants the answer, "What's the plan, Admiral?"

Admiral Whistler apologizes, "Sorry sir, but I thought you might like to hear this part. Admiral Stiller called Paul, the restaurant manager at Maki. Paul is good friends with the Admiral. The Admiral told Paul he was bringing his great friend Colt Jackson back to Maki and if possible, could they be seated near the hostess stand, because Colt kind of liked Akiko and wanted to say hello to her, while we were at dinner."

Lori Sanders entire body lights up with fire. She curtly says, "WHAT?"

Admiral Whistler laughs slightly and says, "It was just a cover, Ms. Sanders. Colt has no interest in Akiko---trust me."

Lori Sanders recovers from her obviously jealous outburst, "I don't care, even if he does." She tries in vain to cover her obvious feelings for Colt.

The Admiral continues, "So, during dinner, Colt makes a big show of himself with the Admiral. They drink and pretend to get a bit too drunk. The Admiral has his guards with him, so there is no risk of another incident with an assassin. But, right in front of the hostess stand, with Akiko all ears, Colt lays out his plan. Remember, when Colt and you entered Maki, Paul greeted him and acknowledged Colt was a Medal of Honor winner---this according to your brief."

Lori Sanders replies, "Yes, I remember, Colt was not happy about that."

The Admiral goes on, "Akiko was VERY nice to Colt. She was smiling, said it was so nice to see him again. She was all cozy with Colt. I guess Admiral Stiller could barely handle it when Akiko came over and put her hand on Colt's shoulder---she's, apparently, a very striking woman. A classic Japanese beauty, I understand."

Lori feels another flush of jealousy rush through her.

The Admiral continues, "And then, just when she is walking away from their table, but in clear ear-shot of Colt, Commander Jackson tells the Admiral, 'Me and SEAL Team 3, all one-hundred-twenty of us, are going to take the Haditha Dam, the night of October 27th. We'll come in by parachute. I want to personally shoot that son of a bitch, Caliph!'"

Haditha Dam

Public domain image. Work product of U.S. Government.

General Shane Maxwell then gives Lori further background, "While you have been busy with everything else, Ms. Sanders, we had a different team follow up on Akiko. It turns out her contact, Hassan, has been making numerous calls from his 'Go' phone to Haditha, Iraq. Further cross checking indicated the Caliph has regrouped there, in preparation for an assault on Baghdad. Interestingly, as soon as Commander Jackson and Admiral Stiller left, Akiko phoned Hassan, again. Immediately after this call, he then placed a call to another disposable cell phone in Haditha. Intelligence, again, confirms this is where the Caliph has moved his headquarters. It is a staunchly Sunni town and is extremely strategic to Iraq."

The Admiral takes it from there, "Commander Jackson was aware of the initial intelligence and developed his plan, accordingly. Getting Hassan to make the immediate follow-up call to Haditha, just cinched it."

Lori Sanders is on the border-line between shock and dismay, she is beyond alarmed, "And that's the plan, one-hundred-twenty Navy SEALs land on the Haditha Dam and kill every single ISIL soldier in town? You are kidding me, right?"

Admiral Steven Whistler confidently tells her, "No, that's pretty much the plan, with a little twist."

The President asks, "What's the twist?"

Admiral Steven Whistler drops the bomb shell, "Colt believes he can be used as the bait. Akiko knows he's a Medal of Honor winner. The Caliph will not be able to resist the prize of either killing, or capturing, a friend of the President's and a friend of Prince Harry's."

General Shane Maxwell jumps right in, "ISIL did their digging on Colt Jackson, after the incident at Maki in Bahrain. Somehow, a part of Colt's file has been leaked to them. We think it is someone in the State Department, or even possibly the White House, but we haven't been able to confirm this, yet."

Lori Sanders asks incredulously, "Colt is using himself as bait? Who is going to kill ISIL? THIS DOES NOT SOUND LIKE A GOOD PLAN, TO ME!"

The Admiral tries to calm Lori down, "Please, Ms. Sanders, let me finish. Colt and his men will land on the dam---ISIL will believe they have them trapped AND SURROUNDED. In fact, they will. Colt and his men will secure the INTERIOR of the dam. It is a fortress and they can hold off any attack which may occur, once they're inside. There will have to be a skirmish on top of the bridge, however, to suck the Caliph and his followers in."

Lori asks, the important question, somewhat sarcastically, "How long can they hold out, once they're barricaded, and, or, trapped, inside the dam?"

Admiral Whistler dismisses her concern out of hand, "The SEALs know how to defend themselves, Ms. Sanders. Each man carries eight thirty-round magazines for his rifle, plus a couple of hundred rounds of link ammo, and three 15-round magazines for their pistols. They also carry hand grenades and rounds for the grenade launchers on their rifles. That's about 500 rounds per man, plus their grenades. Together, they have firepower of over 50,000 rounds of 5.56 mm bullets, which can pierce body armor, and these are being fired from one of the most lethal carbines in history."

Lori Sanders rolls her eyes, "And what would that be?"

Admiral Steven Whistler proudly speaks, "Why, that's the Colt M4A1, with a Special Operations Modification."

Lori shakes her head, "Huh!...of course it has to be called a 'COLT!'"

The Admiral responds, "Afraid so, Ms. Sanders. Its integrated infrared illuminator is good to 300 meters and it has laser targeting, along with night-vision goggles---and you know they'll be wearing their headgear."

Lori Sanders questions, "But, aren't they sitting ducks up there?"

The Admiral answers, "The plan is to only hold the dam for about fifteen minutes. They'll eliminate any of the guards posted, and secure the entrances on both sides of it. It's a big dam---9,064 meters, or 29,738 feet, nearly six-miles wide. It supplies a third of the power in all Iraq and Haditha is a key control city.

From there, ISIL can head north to Mosul and south to Baghdad. Haditha also has a major oil pipeline running through it and the K-3 oil pumping station in town. Aside from being bait, re-obtaining control of the city and this key asset is critical to the future of Iraq. Colt thought we could kill two birds with one stone. He was right."

The President sees the light, "I see, Admiral. So, after the fifteen minute firefight, what next?"

Admiral Whistler smiles, "That's the beauty of Colt's plan. SEAL Team 3 has three 40-man Troops. One will hold each side of the dam, while the other will be spread along the top, providing fire support and making sure no creepy crawlers come up behind the two Troops securing the dam entrances. Once ISIL figures out the SEALs and Colt Jackson are really there, they will pour onto the dam, with overwhelming force."

The President looks at what the battlefield will look like, Lori Sanders puts up a slide showing a picture of the dam. The president asks, "How many ISIL warriors will there be, do you expect."

The General has this answer, "Our latest intelligence shows there are about 20,000 soldiers aggregating in, and around, Haditha. They are getting ready to take Baghdad, we believe. They are also probably planning to blow the dam and have the water crush everything in the Euphrates River Valley---all the way to Baghdad. The dam is mission-critical for the future of Iraq."

The President replies, "I see. Operationally, I leave it up to you men, of course."

Lori Sanders speaks directly to the President, "Mr. President. This is the most asinine plan I have ever heard in my lifetime! In two days, Colt Jackson and a little over one hundred SEALs are going to take the Haditha Dam. What happens when they are barricaded inside the dam?"

Admiral Steven Whistler comes to the rescue of the attack by Lori Sanders on the President. "With ISIL's large, organized force in Haditha, waiting to attack Baghdad, and between the Intel we leaked, as well as Colt's ploy with Akiko, the Jihadists will be literally screaming for Colt's scalp. Destroying SEAL Team 3 would be a great treasure for them, and one the Caliph will not pass up. When they attack, and *after* Colt and the Team are safely under cover inside the dam, we're going to bomb them!"

Lori answers, "But, Colt and his men, you could destroy the dam and they would all die!"

Admiral Steven Whistler smiles around the room, "Not with these kinds of bombs. We're going to use MK77s, a modern-day version of napalm. Each bomb carries 110 gallons of a fuel-gel mix made of kerosene, benzene, polystyrene and white phosphorous. The mixture has an oxidizing agent, making it difficult to put out, once it's on fire. Colt's team will have the doors all barricaded, just before the bombs are let loose. They'll head to the middle of the dam---they will be completely protected in the subfloors of the structure of the dam itself."

"Mr. President, in one fell swoop," General Maxwell says, "Colt believes we can take out the entire ISIL army and, hopefully, the Caliph himself. To make things even more enticing for the Caliph, we let it leak that Colt Jackson and Heaven's Seven were responsible for destroying the cargo vessels out of Amsterdam. From what we understand, this put the Caliph and the Emir into a rage."

Lori Sanders is shaking her head in disbelief, "You are going to napalm the dam and Colt is just going to sit under it and let it happen?"

Admiral Whistler confidently answers, "Yes, ma'am."

Lori is growing more and more skeptical of the "plan," "I don't like it. I don't like this plan one BIT! What did Colt say when he was warning us about the nukes and our dependence on Angel Shield? We were all SO confident. He questioned the most important thing our over confidence failed to do. He questioned our ASSUMPTIONS!"

General Shane Maxwell patronizes Lori Sanders, "Now, don't get all emotional, Ms. Sanders. We've been over this plan with a fine-tooth comb."

Lori Sanders, now furious, says sarcastically, "Of course, sir. What could POSSBILY go wrong?"

The President says, "Everyone calm down. We're all doing our very best here. You're not being emotional, Ms. Sanders. You raise good counter arguments. We need this kind of scrutiny to make the best decisions. However, I think Commander Jackson, the Admiral and the General have created a bold and exciting solution. We never like to use the MK77s, especially in a populated area, but the dam is separated from the town of Haditha. This is a perfect choice of weapon to rid the planet of some horrendously vicious and brutal killers. Look, we're off to a good start here. The SEALs took down the Emir, and I just received a report the 5th Fleet took out the entirety of the Qatari Navy. They also captured Admiral Muhammad Mubarak and their Tomahawk missiles destroyed every aircraft the country owned."

The Admiral nods, "Yes, sir, it really worked. Admiral Stiller just sent a communique indicating the rest of the military was given a choice---surrender, or die. They all put their hands up and ordered their men to disarm. The 5th Fleet and the U.S. Marines are taking over, as we speak."

The President is excited, "Excellent! I think we can all at least be happy about this. Imagine, taking down an entire nation, without losing one of our own. I could just kiss that Colt Jackson!"

Lori Sanders is fuming, knowing Colt is again going into harm's way. "When does the next leg of the mission begin, Admiral?"

The Admiral answers, "Immediately. Colt and SEAL Team 3 are now heading to Bahrain, where they will prepare for the assault on the dam, three days from now. They will pack up for the attack during the evening, of the 27th, and parachute down on the dam during the wee hours of the 28th, Iraq time."

Lori then tries to calm herself down and pull back a bit, "Well, let's just all wish them luck. I hope they don't need it."

Admiral Steven Whistler says, "Don't worry, Ms. Sanders, everything will be all right."

Lori shoots him a dirty, aggravated look.

Chapter 24

Attack On The Haditha Dam

Inside a fleet of Sikorsky UG-60 Black Hawk Helicopters, it is now 08:00AM Qatar Time, 01:00AM Eastern Time on October 25th. The Stealthy Black Hawks are flying the entire SEAL Team 3 up to Iraq. The Black Hawks have a combat radius of 368 miles, or 590 kilometers, with a cruise speed of 150 knots, or 173 miles per hour (278 kilometers/hour).

The journey from Bahrain to Haditha is 1,388 kilometers, or 863 miles. Three refueling stops are planned, along the way. The last one will be at the airport in Baghdad, where the SEALs will be under the protection of armed, United States Marine guard, where they will prepare for the attack on the dam. The Iraqi government has given the United States a wide berth in helping it fight the Islamic State, a proclaimed enemy of the government of Iraq.

The Blackhawk's make their first quick pit-stop in Bahrain, where Colt's team merely reloads its weapons and picks up their swimming gear for the attack on Haditha Dam. They will rest a day when they arrive at Baghdad Airport and get ready to go, late in the evening of the 27th and into the early morning hours of the 28th of October. From Baghdad, it is only 270 kilometers to Haditha and well within the Blackhawks' combat radius.

As the mission begins, Colt's men are well rested. The day of rest in Baghdad was a good one and they were able to play a good game of touch football, with a great steak dinner afterward on the 26th. On the 27th, the SEALs adjusted their clocks and slept in, knowing they were flying into battle during the very early morning of the 28th of October.

The clock always moves slowly when men prepare for battle, and the 27th of October is no different. During the day, everyone on SEAL Team 3 acted like it was just another day at the office, but the bad jokes and attempts to make things normal, seem a bit overdone. All the men are a bit apprehensive, which is perfectly normal before a firefight begins.

Finally, at 03:00AM Baghdad Time October 28th, 08:00PM Eastern Time on October 27th, Commander Colt Jackson's entire SEAL team is airborne in a fleet of Sikorsky UG-60 Stealth Black Hawk Helicopters, heading for the Haditha Dam. Colt looks at his men and can see the look of grim determination on their faces. Each man is going through his own private ritual and preparing his thoughts before the battle begins.

About 30 miles south of Haditha, Colt commands the pilot to make some unplanned changes to the flight plan.

There is only a waning moon in the sky, with a very small crescent, but it is not visible. There is heavy cloud cover, making the night pitch dark. It is perfect for U.S. Navy SEALs to conduct a night operation over hostile territory.

Satellite Imagery Haditha & Dam

Public domain image. United States Central Command.

Colt Jackson orders the lead Black Hawk Pilot, "Lieutenant, I want you to take these vectors to the dam." Colt points to a map with lines drawn, vectoring around the city of Haditha. The revised flight plan is miles wide of the dam. "I want you to keep us out of sight of Haditha, as far as possible, and then bring the stealth choppers a mile-and-a-half north of the dam---right over the middle of the reservoir, Lake Qadisiyah. We're going to swim up to the back of the dam, underwater. Those radical terrorists will all be waiving their AK-47s up in the night sky and we'll surprise them by climbing up onto the back of the dam."

Lieutenant David Preston, the lead pilot, says, "Aye, aye, sir. Our orders are to follow your orders. I'll let the other birds know."

Colt Jackson turns to Heaven's Seven. "Guys, we've really chewed some ground together. Let's see if we can do it again. This one could break the back of ISIL."

The Heaven's Seven aggressively responds to their leader, "HOOYAH, SIR!"

Lieutenant Cody Custer looks at his Commander, "Sir, I think it was a really good idea to not tell anybody about our plan of hitting the drink and swimming, in SCUBA gear, to the dam. I'm afraid of leaks and all the nervousness in the country right now. I'm glad you played this one tight to our chest."

Colt stares Tiger Fox directly in the eyes, "We have to take every precaution we can. We're expected to do dangerous jobs, but God gave us a mind to use it. Tiger Fox, always remember---use everything around you to protect your men---even the people pulling your strings, if you have to."

Tiger Fox replies, "Will do, Papa Bear!"

Colt Jackson barks an order, the second the Jump light turns on, "JUMP LIGHT! Go! Go! Go!" Colt is the last one out.

SEAL Team 3, in its entirety, parachutes from a height of 10,000 feet, with their black parachutes fully deployed. Not one of the 120 chutes is visible on this very dark, very early morning. It is still the middle of the night.

The Team has also activated their Sealskins. After the first 10 seconds, the SEALs mark the 9,000 foot altitude mark. In another 45-seconds, they meet mother Earth.

At a fall-rate of 5-seconds per thousand feet, all 120-men splash down in the reservoir, about a minute after they jumped from their respective Black Hawks. The stealth choppers placed them exactly a mile-and-a-half away from the dam, right in the middle of Haditha Reservoir, otherwise known as Lake Qadisiyah.

On Haditha dam itself, the guards are all looking skyward, but in the wrong direction. They have been leaked information the SEALs will be landing on top of the dam, but *coming* from the south. They received this confirming information from their sympathizers, in both the White House and State Department. The guards point their guns, menacingly, directly up at the sky, looking south from the dam.

They completely miss the ten dozen SEALs splashing down in the lake, a mile-and-a-half, behind them.

On the sloped, concrete wall of the dam which leads down to the top of the lake, it is October 28th at 03:45AM Baghdad Time, and October 27 at 8:45PM, Eastern Time.

When Colt and the Seals prepared for the mission, they went over every detail they could find on the dam. The $830,000,000 Haditha Dam opened in 1987. It is the second largest dam in Iraq, after the Mosul dam, which was built on the Tigris River to the north. Haditha Dam controls the Euphrates and supplies a steady flow of water through Iraq's extensive irrigation systems. It also powers one third of the electricity in the country, through the power station in Haditha. There, six Kaplan turbines generate up to 660 Megawatts of electricity.

Haditha Dam Six-Mile Length

Public domain image. Work product of U.S. Government.

The reservoir behind the dam is called Lake Qadisiyah. It is very large. It covers 193 square miles and has over 60 miles of jagged shoreline, with many coves etched into the lands around it. The dam itself stands 187 feet tall and is an outstanding engineering accomplishment for the country. Its construction was sound for conventional purposes. It was built with asphaltic concrete, detrital dolomites (sandstone consisting of quartz, calcite, clay, rock fragments and translucent minerals made of calcium and magnesium fragments), along with a mixture of sand and gravel. The core is protected by a steel-reinforced, concrete slab.

The dam also contains 16 stories in its structure, as well as base operations. It is from these areas the dam is controlled.

This early morning, at the very top of the dam where ISIL believes SEAL Team 3 will land via parachute, it is lit up by bright street lights. As the SEALs approach the face of the dam underwater, they know the dam has a slope going down to the water, which is an easy uphill climb. With their Sealskins active, they should be able to sneak up on the guards and eliminate them. The only thing visible, in fact, will be their footprints, which will be wet from the water.

Colt and his men know it is important to take down the Caliph, but they also understand the importance of capturing and holding the dam, at all costs. If the dam ever breaks, or is destroyed in war, Baghdad and millions of people in the fertile valley below would suffer and thousands would die.

The first thing Colt will do is send in Recon teams inside the dam. They will determine if it is rigged to blow, if under attack. The SEALs, all 120 of them, are now lined up at the wall of the dam. They lie invisibly at the water's edge, which laps at the sloping wall of the dam before them. Colt delivers his first order over the COMM channel.

"Tiger Fox and Stone Fish, listen up. Take two platoons of men. Thirty-two in all. Our first job is to inspect each level of the dam. There are sixteen of them. Assign two men per floor. Start on opposite sides of each level and work to the middle. Disable any munitions, devices, or booby traps you see along the way. They are bound to have at least some in place. You know the rules of engagement. We are dealing with murderers and brutal fanatics here. Once you've identified friend, or foe, your orders are to shoot first and ask questions later. You have exactly two hours to check the entire structure. Sunrise is at 06:18, we need to move by then. Are we clear?"

Lieutenant Cody Custer and Ensign Dizzy Lee whisper back, "Yes, sir."

Commander Jackson reminds them, "Men, make sure you remind every team, for each floor, to set up their wireless antennae repeaters. I want to have eyes on everything each level sees; otherwise, the dam will block all communications."

With a nod of their heads, they take off.

Inside the interior of Haditha Dam, it is now 04:45AM, Baghdad Time on the 28th, and 09:45PM Eastern Time on the 27th of October.

The thirty-two men all turn off their night vision gear because the interior of the dam is well lit. They dutifully and quickly check the interior of the dam. Each pair of SEALs goes to its assigned level. Lieutenant Cody Custer has taken the dam's lowest level and he has paired himself with

Master Chief Jack Flintlock, AKA "Killer Whale." They are about to meet in the middle, after taking only an hour to check this level of the dam.

Suddenly, Tiger Fox sees wiring which leads to a mountain of C4 explosives. Tiger Fox squawks over the COMM, "Killer Whale, I've found a mountain of C4. I figured if they were going to blow this thing, they would put the explosives on the bottom level. The explosives would weaken the dam most quickly here, if they were ever set off. Are you close? I'm right in the middle, by the machinery."

Killer Whale responds, "I read you, Tiger Fox. I'm about thirty yards away. I'll be there in a minute." Master Chief Jack Flintlock arrives and admires the pile of C4. Killer Whale whistles. "Phewwww."

Tiger Fox orders, "Let's get to work, Chief."

"Roger that. Shall we bring some C4 back up for Papa Bear, you never know?"

"Is the Pope Catholic? Yeah! Let's do it. Make sure you destroy the detonator and let's yank the wires. We don't want to be sitting upstairs and have this baby go off!"

Killer Whale prays his kind of prayer, "Amen to that!"

The virtual video screen in the White House, and the screens at the NSA in Kristi Kruger's and Brian Pierce's computer pod, reveal a mound of C4, rising many feet in the air. There are wires going off in four different directions. The Islamic State has taken no chances. They have four independent detonators on the outside of the dam to make sure this stockpile of explosives goes off, if necessary.

Brian Pierce is studying the mound and listens to the Chief's voice. Brian loses it, "HOLY SHIT! Look at that pile of C4!"

Kristi Kruger is also shocked, "Unreal. That Tiger Fox dude is pretty cool. He's like Colt Jackson's right-hand man. He seems pretty smart, too. He knew where to find the explosives and took the most dangerous floor---typical Colt Jackson stuff. These guys are a dream."

Brian Pierce sighs---a sigh of frustration and dejection.

Inside the White House Situation Room, the observers are wide-eyed. The scene on the video wall was like some of the ammunition caches discovered in Baghdad after Saddam Hussein was overthrown. Apparently, they didn't find them all. This pile of C4 rose no less than ten feet tall.

Admiral Steven Whistler states the now obvious, "I would say Commander Jackson's instincts were right, once again. Can you imagine if they set that off? THOUSANDS would die from a massive and violent wall of water and Baghdad would be destroyed."

The President looks at the mound of explosives and says, "I shudder to think. Holy smokes."

General Shane Maxwell looks on and says, "Thank God they found it."

Lori Sanders agrees, "You can say that again. WOW."

Deep within the bowels of Haditha Dam, the men destroy the detonator and pull the wires off the C4. Suddenly, over the noise of the equipment inside the dam, they hear the loud retort of an AK-47. Killer Whale takes the shots to the back of his body. He falls over, lifeless.

Killer Whale cries out, "Ugggh!"

Tiger Fox spins and grabs his MK 23 H&K .45 caliber pistol and fires while he twirls. The Islamic guard's forehead explodes from the impact of the large caliber .45mm bullet. He had snuck up behind them while they were working on the C4. Lieutenant Cody Custer rushes to the Chief, who is lying motionless, before him.

"Chief, Chief, are you OK?? CHIEF!!" Cody feels for a pulse and then shakes him."

Killer Whale begins to come to. "Uhhh…Tiger Fox. It knocked the wind out of me and I think I took a round to the back of my headgear because my electronics are acting weird. I think I blacked out."

"Thank God you're OK. I don't know if there are other guards down here, but we need to high-step it back upstairs. The sooner we get back, the better. At least the machinery covered the sound of the gunfire.

"Let's go, sir. I'll be OK. What should we do with the body?"

Tiger Fox looks around. "Let's stuff him under the C4!"

Killer Whale replies, "Now that's a fine idea. Perfect!"

At Kristi Kruger and Brian Pierce's computer pod, they both breathe a sigh of relief---Killer Whale is OK.

Kristi romantically says, "See what I mean? He's just so cool! What a way to send a murderer to Allah---you send him in a million different directions if the C4 goes off! Tiger Fox is quite a guy!"

Brian Pierce rolls his eyes and continues watching the screens before him.

Inside the White House Situation Room everyone also breathes a sigh of relief.

On sloped concrete, near the water on the back of Haditha Dam, it is now 05:15AM Baghdad Time on the 28th, 10:15PM Eastern Time on the 27th, in Washington.

Lieutenant Cody Custer and Master Chief Jack Flintlock are the last two SEALs to return to the top of the dam. They turn their UV and infrared on and now see scores of SEALs around them. As they look, in both directions of the dam, they can see ghost-like, glowing bodies lined up, all the way across it and red-hot bodies patrolling across the top. Those were the guards.

At the far side, another cluster of men is poised to attack the guards on top. The Islamic guards on the dam have been constantly monitoring the night sky. Not one sees the SEALs with their active camouflage turned on and they are only a matter of feet away from them, lying at the lip of the dam, below.

Lieutenant Cody Custer reports to Commander Colt Jackson and gives his Commander a sit-rep.

"Hello, Papa Bear, it's Tiger Fox. Over."

Suddenly, Tiger Fox is poked in the back. Tiger Fox spins, but sees it is Papa Bear, who had worked his way over to Tiger Fox's side of the dam.

Papa Bear greets him, "Hello, there, Lieutenant. No need for mikes now. Everything OK?"

Tiger Fox replies, "As I'm sure you could see in the Seaglass, sir, we disabled a mountain of C4 and killed one guard. The guard got off multiple AK-47 rounds on Killer Whale, fortunately, they hit his body armor in the back, but he also took a round to his helmet."

Papa Bear turns to Killer Whale, "I saw that, but also saw you got up. Are you OK, Chief?"

"AOK, Skipper. I guess I have a thick head."

Papa Bear jumps on that one, for the audience back in Washington---for one person, particularly, "Yeah, I've been told that, too."

Inside the White House Situation Room the evening is getting later. It is October 27th at 10:20PM Eastern Time, which is October 28th, at 05:20AM Baghdad Time.

Lori Sanders is listening to Colt's conversation with Killer and can see both men's faces on the screen before her. When Colt makes the remark about being thick headed, the room breaks out in laughter.

Lori Sanders speaks up, and speaks some "French," "What a wise ass! And please excuse my French. He's doing that on purpose! He KNOWS we're listening to everything he says. Sometimes, I'd like to..." She bites her tongue, smiles and goes along with the laugh.

Admiral Steven Whistler tells Lori, "He does have a sense of humor---you have to give him that."

Lori is still smiling because she knows that comment was intended for her. "*He's still thinking about me, at least.*" she thinks. She responds to the Admiral, "Thank God the Chief is all right." She pauses and cocks her head toward the Bose speaker system, "Listen. Does anyone hear that?"

The President perks up, "Hear what?"

General Shane Maxwell replies, "It sounds like the wind is picking up."

Lori Sanders starts tapping on her computer keyboard. Her eyes grow wide and she utters, "UH-OHHHH! It looks like a storm with a strong cold front is passing over the mountains of Turkey. Could be a windstorm kicking up. It's going to be hot there today also. It will be over 100° Fahrenheit."

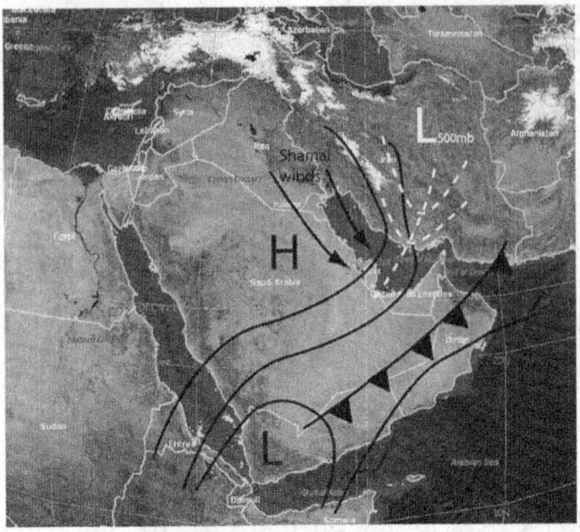

Shamal Winds

Map based on NASA Worldwind. The meteorolgical conditions are based on "Wind regime of the Arabian Gulf" by Ali Hamid Ali published in the book The Gulf War and the Environment, By Farouk El-Baz, R. M. Makharita, Published by Taylor & Francis, 1994, ISBN 2881246494, 9782881246494. Date: 16 January 2009. Own work of author, Pcirrus. http://commons.wikimedia.org/wiki/File:Shamal1_en.jpg

The President looks at Ms. Sanders, "Please speak English for me, Ms. Sanders."

"Sir, if a cold front moves in, the leading edge of a mass of cooler air causes dust and sand to kick up. It sends them in the air and can create those epic sandstorms you sometimes see on TV. You know, like the ones we have seen hit Phoenix from time to time. There is something biblical about them. In the Middle East, they call the thick wall of dust and sand "Haboob." In Iraq, it means simply, "Strong Wind."

Admiral Steven Whistler is looking worriedly at General Shane Maxwell. He states, answering the General's question, before the General can even open his mouth, "We can't pull the plug now, it's too late. They're going to start firing any minute."

General Maxwell thinks and then says, "There isn't a sandstorm yet. We will have to deal with that, when and if, the time comes. One thing for sure---we can't allow ISIL to hold that dam! Who knows when they might blow it with C4! This is NOT a mission about one man---it is a mission about our country and the stability of an entire region of the world. The SEALs will do their job. It's time to stop worrying they won't, Ms. Sanders."

"Sir, I wasn't worrying, I just picked up on the wind noise. Somebody should at least communicate this Intel to Commander Jackson."

Admiral Steven Whistler responds, "That is my call. I will if the weather conditions worsen. Until then, let them focus on their job. They have to take out all the guards on a synchronized basis. Colt wanted to try and eliminate the threat of destruction to the dam first, then eliminate the roving guards across six miles of dam, second, and accomplish all of this, without being caught. At daybreak, he suspected all hell would break lose when ISIL figured out SEAL Team 3 held the dam. They'll be pissed, especially because they were tipped off in advance by three different sources: Akiko, and two other sympathizers within our own government!"

The President inquires, "What are you doing to catch these traitors?"

General Maxwell answers, "Everything sir. We already tracked down their disposable phones and where they were purchased. It seems they used some unemployed people to make their purchases for them. We've already interviewed several of them. We have descriptions, but nothing on video, yet."

The President says, "Keep me posted on this. I want to route them out. They deserve to be hung."

General Maxwell responds, "Yes, sir, we will."

On the backside of Haditha Dam, on the sloped concrete, near the water at 05:28AM Baghdad Time and 10:28PM Eastern Time, Colt speaks into his mike to all the men. They receive his orders on their ear buds, over the encrypted SEAL network. Everyone can read the digital read-out on their headgear. It clearly shows the countdown-time to execution.

When the clock ticks down to "0.00" one-hundred-twenty SEALs, currently strung out across six miles of dam, will take out three-hundred armed guards. The guards are dressed in their black Bisht, Islamic garb and are patrolling the six-mile length of the damn. This amounts to one guard for every 106 feet of dam. As Colt Jackson and SEAL Team 3 assesses their targets, the guards are all still looking up at the sky.

"Gentlemen, on the mark, take out your targets. You'll have to pivot and take out your next pre-assigned targets in your kill zones, just like we planned. Here we go."

The clock strikes "0.00". One hundred SEALs, armed with Colt M4A1 carbine rifles, and twenty SEALs armed with FN MK20 sniper rifles, open fire with suppressed weapons. The sound emitted is eerie.

SEAL Sniper Rifle Haditha Dam

Public domain image. Work product U.S. Government.

Inside the White House Situation Room, the President says, "It sounds like a huge gust of wind blew for exactly a half-a-second."

Immediately, one-hundred-twenty bodies drop, scattered evenly across the bridge. They collapse as dead weight. Guards across the dam see their now dead brothers. They immediately look up at the sky. The Islamic guards hear the collective "WHOOSH" of the silenced weapons, but can't tell from what direction the sound is coming from, because the real wind is now howling and dust is blowing through the air.

The dust hasn't affected the SEALs because their night vision optics on their Colt M4A1s and their sniper rifles work perfectly.

The one-hundred-twenty men of SEAL Team 3 go about their business. They set their laser sights and lock on to their next targets. Prior to firing, each SEAL calms himself by taking a deep breath and then gently squeezes his trigger.

The guards are now menacingly alert and point their AK-47s at the sky and begin to pull their triggers. Just before they squeeze them, another one-hundred-twenty bodies drop. Eighty-three of these trigger pulls, were headshots. Upon impact, brains and tissue mass explode out the other side of these guards' bodies. Unfortunately, while the next one-hundred-twenty men fall, a roar of gunfire goes off into the sky.

Eighty more ISIS guards drop. At least half of them also experience the entry of a bullet on one side of their brain, and the exit of that bullet out the other side of their brain. 41 more heads explode. This grotesque site is seen live at The White House and back at the NSA.

At Kristi Kruger and Brian Pierce's computer pod, Kristi turns to Brian and says, "I can't watch that." She shields her eyes.

Inside the White House Situation Room at 10:29PM on the 27th, Eastern Time, and 05:29AM Baghdad Time on the 28th, Admiral Whistler immediately comments when he hears the roar of the AK-47s fire off. "Uh-ohhh, here we go! They're going to be in the thick of battle real soon."

Colt Jackson takes command of his men, "OK, men, until we know for sure, the ISIS force is *primarily* on the west side of the river. Troop Alpha, which has created a perimeter defense on the far-east side, will remain in place. Troop Bravo---head over to Troop Charlie where I am, just to the west of the middle of the dam. We need to reinforce this area---quickly. This is where the key entrances are and this is where the critical infrastructure is for the dam. It's a trot, I know. Some of you may have a couple of klicks to run, but do the best you can."

Multiple squawks are immediately heralded across the Navy SEALs' COMM line, in reply. Troop Bravo immediately pushes up from their firing positions and runs directly west---towards Papa Bear, Heaven's Seven and the rest of Troop Charlie.

The Situation Room has the ability to switch on, and off, various video cameras being broadcast from any of the one-hundred-twenty Navy SEALs deployed from SEAL Team 3. Three of the large viewing screens are split into forty different streams, creating miniature-sized video walls.

These screens serve a special purpose. They allow an overall, simultaneous view, of every member of SEAL Team 3. In effect, each individual stream allows each soldier to record his own battle. The rest of the wall, appropriately, shows the Troop, platoon and squad leaders' viewpoints. The system is extremely robust and can be interchanged at will, with any of the screens moving to the action, in an instant. Currently, the video wall for Troop Bravo, displays video motion and running movement from forty different cameras---all heading directly west to reinforce Colt Jackson and Troop Charlie.

At the west-side security perimeter in the middle of Haditha Dam, Colt checks his watch. It is now 05:30AM and the wind is getting considerably worse. Colt looks toward the east. The sun will rise in 48 minutes and the sky is already brightening. Sunshine is not a SEAL's best friend. It evens the playing field for the Islamic fighters. Colt knows why there hasn't been any gunfire, yet, from either the western or eastern sides of the dam. The sound of a massive volley of gunfire from the guards, just before the last group died, would have been easily heard in the town, below.

Lori Sanders asks the Situation Room, "Why isn't ISIL attacking? Anyone could hear those AK's go off---all the way down the valley."

People not in the direct chain-of-command can "see," but cannot "speak," to the soldier, during a battle. This ensures SEALs are not distracted by others during the strains and rigors of warfare and this includes the civilian employees of the NSA.

Colt warns his men, "Gentlemen, everyone, listen up. We are up against a very clever enemy. The Caliph has a PhD and twenty years of battlefield experience. He prides himself on being a Prophet and follower of Muhammad."

Ensign Dizzy Lee, echoing Lori Sander's unheard question asks, "Why aren't they attacking, sir? They must have heard the shots."

Inside the White House Situation Room, The President echoes the sentiment, "Yes, why isn't ISIL attacking, yet. They must have our boys outnumbered thousands to one." The President looks over at Lori Sanders. Lori looks down at her computer, feeling a sense of dread fill her body.

Tiger Fox answers Stone Fish, "It's because of our night vision gear. They know we're here now. They probably have no idea how we got here, but nevertheless, they know we're SEAL Team 3. We're the bait."

Commander Jackson affirms what Lieutenant Cody Custer just told Ensign Dizzy Lee, AKA "Stone Fish." "Exactly, Tiger Fox. They won't attack us at night, when we control the top of the dam. They know we *own* the night. If we had parachuted down on them, directly from above, instead of landing in the lake and sneaking up behind, they would have used their roving patrols to engage us---despite our advantage at nighttime. Now, however, they will wait for sunrise. They will also stay hidden. Remain alert. I guarantee you there are thousands of eyeballs on us, right now!"

Chief Spider Sinclair asks, "Sir, what about this wind, it's really getting strong."

Colt answers, "I know. We have to wait until dawn to see how bad it is out here. For now, leave your night vision on. Stay alert."

Inside the White House Situation Room, Lori Sanders is unconsciously rubbing the back of her index finger against her mouth. She's definitely getting nervous.

Over at the NSA, Kristi Kruger and Brian Pierce continue observing. Kristi, witnessing the action before her, begins to speak, "He's…"

But, before she can speak, Brian cuts her off. "I know, I know. Tiger Fox is Mr. Wonderful."

Kristi corrects him, "Well actually, I was going to say Commander Jackson looks pretty hot, too!"

Brian responds, "I think I'm going to be sick. I'm going to be Lori right now, OK? GET BACK TO WORK!!" He smiles at her, albeit with a hurtful expression on his face.

"Brian, COME ON. They're hot, admit it. But, when this is all over, I have to ask you a question. But, not right now…"

"OK, OK...they're hot---I mean if you like guys, and, by the way, you can ask me a question any time you want, no problem." He smiles at her. She smiles back. Brian takes her in and fully realizes how beautiful and brilliant she is.

Kristi Kruger truly has an incredible smile, deep, chocolate-brown eyes and beautiful, wavy, long-brown hair. Brian can only dream about her.

Troop Charlie is holding down the fort on the west-side security perimeter for Haditha Dam. It is approaching dawn. In Baghdad, it is October 28th at 06:00AM and it is October 27th at 11:00PM Eastern Time in Washington.

The loudspeaker from the main mosque in Haditha comes to life with the call to prayer. The SEALs look down on the valley and can see the infrared images of many, many people. Even at this early hour at dawn's first light, they come out to pray. They all face east, towards Mecca. If the SEALs didn't have their infrared goggles on, they wouldn't be able to see below.

The wind is really racing now and blowing over 25 miles per hour. It is growing stronger by the minute.

Colt Jackson orders Lieutenant Cody Custer, "Tiger Fox, spread the word. We may be in for a 7th Calvary stand if the wind picks up any further. Sandstorms can begin when the wind hits about 25 miles per hour. I bet we're almost there. If it gets any worse, expect one of those titanic waves of dirt and sand to engulf us from the northwest---they come down from Turkey in this part of the world. They also can tower over 10,000 feet and cover hundreds of miles."

Tiger Fox replies, "Roger that, sir."

Kristi Kruger didn't ever really study history.

She innocently asks Brian, "This isn't the question I told you I wanted to ask you, but what is a '7th Calvary stand?'"

Brian Pierce replies. He is an avid student of history. When he's not tinkering with computers, he is a voracious reader. "On June 25, 1876, General George Armstrong Custer, leader of the 7th Calvary, was butchered. He was completely wiped out by a combination of the Lakota Sioux, the Cheyenne and Arapaho Indians, led by Crazy Horse and inspired by Sitting Bull. Two-hundred-sixty-eight men died that day, including every man in Custer's troop. Custer had also split his brigade into three separate troops, just like Colt Jackson has. Separated, over ten thousand Indians attacked. While the men fought bravely, they were completely outnumbered and were wiped out. Everyone was scalped, except General Custer, for whom the Indians actually had great respect. At least that is how the story goes. There has been a lot of romanticizing it over the years, but I believe this is pretty much the truth."

Kristi looks at Brian, apprises his answer and says, "So, let's see, Colt Jackson has a total of 120 men, they're sitting on a narrow wall, surrounded by water and probably at least 10,000 screaming fanatics ready to come at him. I think I know now, what Commander Jackson is talking about. THANK YOU, BRIAN. I REALLY MEAN IT. It is easy to sit and look at a screen and somehow detach yourself from reality. These are men in what could be a terribly desperate situation---especially if aircraft get grounded by this wind storm we're starting to hear about!"

Brian answers, "Kristi, don't get me wrong, but I just want to pay you a compliment. You are not only *beautiful*, but your mind never ceases to amaze me. You have heart. I love that about you."

Kristi perks up and replies, "Thanks, Brian! By the way, isn't that guy, Tiger Fox's last name Custer, also? Let's hope it's not a bad omen." She smiles at him and raises her eyebrows in concern, but is trying to be positive. "Let's hope for the best these guys kick butt. Hopefully, they'll get the support they need."

Inside the White House Situation Room, Lori Sanders begins to feel her temperature rise. She is getting a very bad feeling inside. The Admiral, Lori notices, is fiddling with his tie. General Shane Maxwell takes a handkerchief out of his pocket and folds it, again and again, for no apparent reason. The President's jaw is clenched, tight as a drum. You can feel the tension in the room---it's as sharp as a razor.

The President asks, "How far are they from the Baghdad Airport?"

The Admiral knows this distance down to the centimeter, "They're 252.7 kilometers, or 157 miles, sir."

The President asks him, "Can you launch the aircraft now?"

The Admiral replies, "We certainly could sir, but ISIL hasn't shown itself yet. They're being very clever. We heard our men talking about ISIL avoiding detection because of the SEALs' advantage with their night vision gear."

The President is deep in thought. Concern spreads all over his face, "Hmmm."

By 06:30AM on the 28th on the west-side security perimeter in the middle of Haditha Dam, Colt Jackson and his forty-man Troop are still, literally holding the fort.

The sun is fully raised and the temperature rises quickly. The clash between the cold front and the warm, blazing sun causes the wind to become even stronger. An enormous cloud of dust can be clearly seen on the horizon---coming toward SEAL Team 3. It will completely cover

Haditha and the Haditha Dam. It seems to go from horizon, to horizon.

There will be no help for Colt Jackson and SEAL Team 3.

Colt looks below the dam, into Haditha. He can see the Islamic warriors gathering on the western side of the dam. He clicks his mike reaching out to the east side of the dam to the leader of Alpha Troop, Lieutenant Michael Perry, call sign, "Speedy Rabbit".

"Speedy Rabbit, this is Papa Bear, do you have any movement to the east?

Lieutenant Michael Perry replies, "Negative, Alpha Bear. Night vision is still effective through the dust, but there has been no movement."

Inside the White House Situation Room at 11:35PM Eastern Time on the 27th, Colt Jackson's video camera is still transmitting in the infrared spectrum. It reveals a terrifying site. Colt has climbed up a structure to get a better view of the enemies' position.

Below the dam, on the west side, Colt Jackson sees no less than 20,000 angry, cheering Islamic Jihadists waving their AK-47s in the air. Everyone in the White House Situation Room, as well as back at the NSA, can hear the ghastly and fearful chant echoing up, from the valley floor, below.

On the video wall in the Situation Room and on the screens of the NSA, the Islamic Jihadists are piped in, loudly and clearly, over the built-in Bose speaker systems:

أكـــبر الله! أكـــبر الله! أكـــبر الله!

لأمريكـــا المـوت!

ALLAHU AKBAR! ALLAHU AKBAR! ALLAHU AKBAR! AL MOUT LI AMREEKA!!

(Allah is the greatest! Allah is the greatest! Allah is the greatest! Death to America!).

One of the president's staff watching the thousands of terrorists on the screen asks, "I know what Allahu Akbar! is, but what does the "Al Mout Li Amreeka!!" mean?

The President, despondently says, "It means, 'Death to America.'"

Silence fills the Situation Room. The tension is palpable. It is obvious what is about to happen.

Lori Sanders explains to the room, "They're waiting for the sandstorm to strike. They are going to slaughter them! She rushes from the room---I'll be right back, I need to phone Kristi and Brian."

Around the security perimeter in the middle of Haditha Dam, it is now October 28th at 06:40AM Baghdad Time and October 27th at 11:40PM Eastern Time.

The sandstorm sweeps over Colt and the entire SEAL Team. They stay focused on the approach of the terrorists. One of the things Colt had his men do, when they swept the dam's 16 levels, was to set shaped charges on any doors which might provide an entrance into the belly of the dam. The men had rigged every one of them. If they were opened, their entrances would collapse. This way, Colt and the Team didn't have to keep looking over their shoulders from within the dam, itself. They would have enough to do protecting themselves on all sides, with their backs against the water. The last thing they needed was to have an army of ants stream up from below.

Lieutenant Cody Custer reflects on this aspect of Commander Jackson's strategic defensive orders, "Good thinking on rigging those doors, sir. If they try to enter, the entrances will collapse. We should be OK on our six, below."

"Great work on making it all happen Lieutenant. By the way, if anything happens, I just want you to know you're one of the very best. I mean, among the best of the best. You've always been there for me and for the team."

Lieutenant Cody Custer looks his commanding officer right in the eye and says, "Right back at you, sir."

Papa Bear gives his instruction, "SEAL Team 3, get ready to open fire, we are going to become a major war zone. Make your shots count. We don't know how long this sandstorm will last and one thing is for sure---they will not fly those jets through the dirt and sand. Their engines will stall and they'll crash and burn. This storm is encompassing everything from here to Baghdad, that's for sure. Look!"

The entire SEAL Team takes a moment to behold a truly Biblical sight. The sandstorm is nearly black from all the dirt in the air---truly a black storm. The SEALs, they now realize, are on their own. There will be no strike of FA/18 Hornets with MK77 incendiary devices to burn their enemies to death, while they hide in the safe confines of the dam, below.

The Jihadists have now massed along the edge of the Euphrates, right near the middle of the dam.

Colt knows any Tomahawk strike, using GPS, can also be ruled out. ISIL is massing too close to the dam. Any strike could easily damage its earthen-sloping walls, creating and causing a great catastrophe. Colt thinks, "*The Jihadists aren't stupid, they know what they are doing. They are massing where the key entrances to the dam are.*"

Colt's decision to establish a defensive perimeter along the top of the dam, protecting its core facilities, turns out to be the best thing he could have done, given the very grim situation quickly taking place before him and his men.

Before leaving on the mission, Colt had read the report from Stuart Bowen, from the Office of the Special Inspector General for Iraq Reconstruction ("SIGIR").

His report on the Haditha Dam found enormous issues with perimeter security. The report, issued April 13, 2010, noted enormous sections of the perimeter fence which did not conform to contract requirements.

As a result, there are many headaches involved with trying to provide perimeter security along the six-mile long dam. These include concrete anchorages for the fencing that were supposed to be 45 inches deep, are only sunk into the desert sand 12 inches. As a result, there are large gaps between the shallow anchorages and

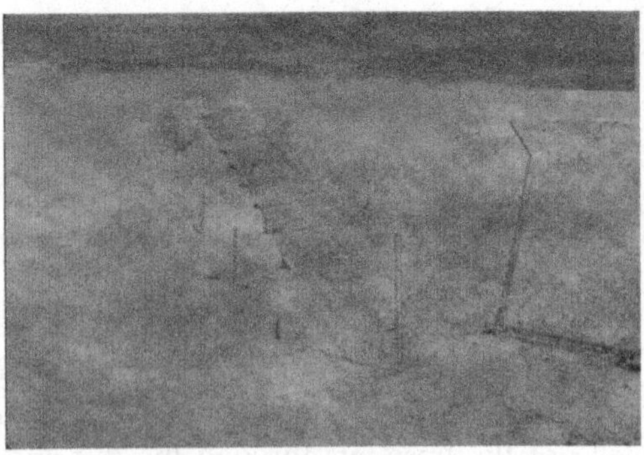

Security Fence Toppled Over
Sandstorm, Haditha Dam.
Public domain image. Work product of U.S. Government. Artist affect: Robert L. Kelly

the now eroded soil around them. Strands of barbed wire are supposed to be strung along the top of the fence, but are not even placed in their brackets, and to put the icing on the cake, large sections of the security fence are completely toppled over.

Between ISIL's RPGs, mortars and other artillery, along with the dam's gentle, sloping rise to the road on top, it was all but impossible to protect the entire length of the dam from attack.

Colt knew the only thing he and his SEALs could do, would be to hunker down and prepare for a tremendous firefight against overwhelming forces---and hope the sandstorm stops and the cavalry shows up.

This is SEAL Team 3's only hope for survival and every man knows it. They were hopelessly outnumbered and if the SEALs scramble into the interior of the dam without the FA/18 Super Hornet MK77 incendiary weapons supporting them, they will be sitting ducks inside.

With Haditha's K3 oil-pumping station close by, and a huge water reservoir at hand, the Jihadists would only love to either burn, or drown the SEALs out, like rats. All they had to do was find a pump, or have the twenty thousand Jihadists pour buckets of oil down into the interior of the dam and light them on fire.

Colt realizes the problem. He has no choice but to fall back from the Jersey security barriers, designed to prevent car bombs, and protect the heart of the dam. In fact, the only advantage SEAL Team 3 has now against the Jihadists, are the Sealskins' cloaking capabilities and their physically higher position on top of the dam.

These advantages are countered, of course, by the perimeter security issues and the disadvantage of the gently-sloping face on the dam itself. The gradual, dirt slope would make it easy for ISIL to breach the dam's perimeter wall and enter the roadway, anywhere along its six-mile length. This could allow the Caliph to surround the SEALs with an overwhelming force and put Colt Jackson and SEAL Team 3 in a very bad position. Adding to these major issues is the power of the black sandstorm. It is preventing any kind of help from the air.

Suddenly, Colt realizes he may have another very big problem.

Suddenly, so does the White House.

Colt Jackson is standing next to Tiger Fox, Lieutenant Cody Custer. "Lieutenant, does your infrared look a little splotchy and blurry?"

Tiger Fox, "Affirmative, sir."

Haditha Dam's Gentle Slope
On a Clear Day & During the Black Storm

A clear day & with a sandstorm. Public domain image. Work product of U.S. Government.

"Do me a favor, turn it off."

Both Tiger Fox and Papa Bear turn their infrared off.
Inside the White House Situation Room, it is 11:50PM Eastern Time, October 27th and 06:50AM Baghdad Time on the 28th.

Lori Sanders gasps in horror. "OH MY GOD! THEIR SEAL SKINS ARE COVERED WITH SAND AND DIRT. THE ACTIVE CAMOUFLAGE DOESN'T WORK IN THE SANDSTORM!"

The President utters, "Oh, Lord, God help them."

Admiral Whistler says to himself, "*Dig in, Colt, use the perimeter wall of the road as a defensive barrier.*"

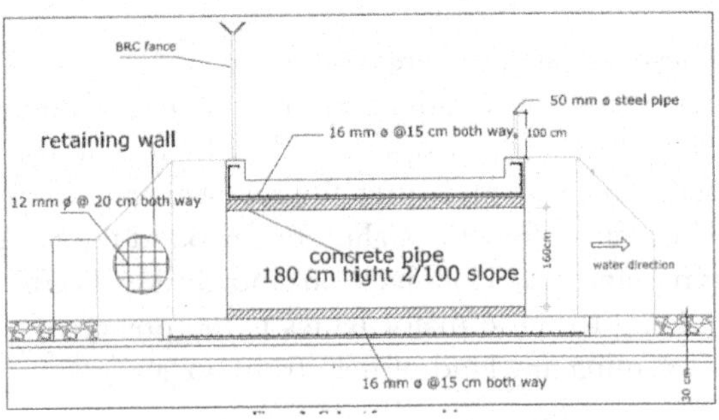

Perimeter Access Road Cross Section
Office of the Inspector General For Iraq Reconstruction
Haditha Dam Perimeter Security, Sustainment Assessment
SIGIR-PA-09-192, April 13, 2010, Page 5.

On the west-side security perimeter in the middle of Haditha Dam, Colt Jackson orders, "Tiger Fox, get on the COMM. Tell the men, the active camouflage---the cloaking capability, doesn't work in the sandstorm. The coating material is supposed to bend the light around us, but it is being plugged up by sand and soot. Its light-bending abilities have been destroyed by the elements. Right about now, ISIL will be able to make out ghost-like figures along the top of the dam. All hell is about to break loose! Get ready and tell the men. We need to treat this just like Fort Apache. Use the fence along the perimeter for protection. Tell the men NO ONE goes into the interior of the dam for protection---if we take cover there, they can either smoke us out, or drown us. We're not going to hide in a burrow. We're going to attack! Tell the men to get ready, on my mark."

Tiger Fox's eyes sparkle with a brave, fierce stare, "HOOYAH, SIR!"

Back at the NSA inside Kristi Kruger's and Brian Pierce's computer pod, Brian is the first to speak, "HOLY CRAP!"

Kristi understands the grim reality of the situation. She could be witnessing a complete wipe out of an entire SEAL Team---never done before in the history of the United States.

She says out loud, "Oh my good Lord in heaven. Heaven help them. They are in deep shit now."

Inside the White House Situation Room, Lori Sanders is praying out lout, "Oh, please, God. Help them."

Back on the west-side security perimeter in the middle of Haditha Dam the clock is at 06:55AM Baghdad Time on the 28th day of October and in D.C., it is 11:55AM Eastern Time, still the 27th.

Colt Jackson squawks over his encrypted COMM line, "Speedy Rabbit, this is Papa Bear. It looks like we are back at the Little Big Horn. I'm not going to have my troop split up when we're facing the entirety of the Indian nations, like Custer did. Get your butts over to Troop Charlie's location, we're going to engage and we'll be under heavy attack. We'll need your support."

Lieutenant Michael Perry immediately COMM's back, "Papa Bear, this is Speedy Rabbit. We're sprinting to you, sir. We'll be there in 15 minutes."

Colt Jackson responds seriously, but in an otherly-world calm voice, "This could be over in 15 minutes if they swarm and use mortars. Do the best you can."

Speedy Rabbit answers, "I'll be there in 12, sir."

Inside the White House Situation Room, the calendar has now turned over to the same day as it is in Iraq. It is October 28 at 12:00 Midnight Eastern Time. It is 07:00AM, Baghdad Time.

The President turns to the Admiral, "Isn't there anything we can do?"

The Admiral replies, "If you authorize the use of a Tomahawk strike, we can pulverize them, but run a nearly 100% risk the dam will break."

General Maxwell enforces what Admiral Whistler has just told the President. "Mr. President, this would kill tens of thousands of people."

Lori Sanders gulps and adds, "When we analyzed the dynamics of the infrastructure of the dam and the potential for sloppy work in its construction---it was engineered by a Turkish company---the odds were very high of a breach. The dam has had major problems with irrigation and, as you can see from this series of photos, the security, drainage and overall design of the dam is questionable if it is placed under the duress of an explosive missile attack. It is simply unable to withstand a high-explosive attack. The MK77s would have worked, because they create intense heat and fire. Sandstone is unaffected by those elements.

The President exhales in turmoil. "Ohhhhh, nooooo....*Assumptions*, you were right, Ms. Sanders. We didn't include this in the threat assessment."

"Mr. President, it's not about being right. Once we made the decision, I supported you, the Admiral, the General and Commander Jackson 1000%."

General Maxwell, "We know it, Lori. This is going to be painful for everyone. There are 20,000 radical Islamists down there and they can access the dam, from any point, along a six-mile perimeter. Commander Jackson and his men will likely be surrounded, on all sides. The Caliph will undoubtedly send a strong contingent to also over the east side of the Euphrates. Once they are in place, the Caliph and ISIL will close their grip around Commander Jackson's neck. The SEALs will be caught in a withering cross fire."

The President looks at Ms. Sanders, "You don't have to watch this. It could be ugly."

Lori Sanders determinedly responds, "Mr. President. I am a key member of the NSA and I run PRISM for Internet operations. I am *also* involved with Commander Jackson, on at least a highly personal, yet professional level. If he is going to die today, I *will* be there with him *and* with you, doing everything in my power to help. I just need a moment, if you don't mind."

Tears are rolling down Lori Sander's cheeks.

The President answers resolutely, "Of course, Ms. Sanders, please, take as much time as you need. Your seat is waiting for you."

Lori answers, "I will be right back." Lori departs for the ladies' room. It is located down the hallway from the Situation Room. She must hurry.

It is now the 28th of October at 12:05AM, Eastern Time, and 07:05AM Baghdad Time.

Lori Sanders has for the second time, left the Situation Room. The first time, she had claimed she needed to speak with Kristi Kruger and Brian Pierce. The second time, she said she needed the ladies room. The truth was, what she was really doing was committing treason, by divulging national security secrets to un-cleared people.

Lori Sanders again dials her cell phone number.

Commander James Forrestal answers the encrypted call with no caller ID, immediately, "Yes, Ms. Sanders, what's the sit-rep?"

Lori responds, "It's a complete disaster. Now their Sealskins don't work. The sandstorm has damaged the coating and light-bending materials. They are sitting ducks on top of the dam and they positioned right in the middle, defending the heart of the dam's operations. Colt has them dug in. They have built fox holes in the aggregate roadway---it is built out of rock and dirt from the desert. This will help them avoid shrapnel if and when mortars come down. The real problem is being overrun by 20,000 screaming, fanatical Islamic Jihadists. They are armed to the teeth! Could you do anything?"

Commander James Forrestal replies, "I'm not sure, ma'am. All I can say is, everyone is trying. It doesn't look good. The storm is playing havoc with everything---communications, machinery, equipment, aircraft---EVERYTHING!"

Lori Sanders cries into the phone. "I have to get back, thank you for trying. And thank you for picking those men up in the Caribbean."

Commander James Forrestal replies, "Ma'am, thank you for leaving me out of the report. Colt told me you nearly figured everything out. He's a good man and I know you did your job, to the letter, but actually saved our butts by holding out certain key pieces of information. Thank you. I know Colt was grateful, also. He told me you were a great woman."

Lori Sanders responds, "HE DID?" Lori is relieved that even after their fighting, Colt feels this way about her. This is some comfort to her, as her love faces certain death.

Commander James Forrestal answers directly and surely, "Yes, ma'am!"

Lori replies, "Commander, I have to go. IF ANY NEWS HAPPENS, CALL ME AT ONCE."

Commander James Forrestal replies, "Yes, ma'am. God help them. Good bye."

Lori reenters the room. It is now 12:09AM Eastern Time, 07:09AM Baghdad Time, her eyes are wet. She asks, "What is happening?:

On the west-side security perimeter in the middle of Haditha Dam, Commander Colt Jackson speaks clearly into the COMM, "Men on my mark, let's give them one big HOOYAH! 3...2...1...FIRE, OPEN FIRE!!"

Eighty U.S. Navy SEALs open fire on the enemy massing below. They mow down hundreds of Jihadists. The Jihadist fighters scramble for cover, but have not volleyed a shot---yet. They are waiting for their orders. The Jihadists are highly organized and are broken down into leadership cells. The money they have obtained, and people they are supported by, has greatly enhanced their fighting and organizational skills on the battlefield.

Each commander is wearing an ear piece and has a headphone strapped around his cotton Keffiyah---the common Arab headwear which protects them from the sun and sand. The SEALs' positions are fully revealed now, thanks to the sandstorm destroying the active camouflage they have been wearing. They are no longer cloaked.

At the western access gate of the Haditha Dam, Caliph Abu-Bakr is speaking over his microphone, "Listen, EVERYONE! Allah has blessed us, indeed. He has used the sandstorm to reveal the SEALs' position! Rain fire on them!! Allahu Akbar!"

Upon his command, nearly twenty thousand ISIL fighters open fire on the SEALs' positions. The massive volley of raw fire power forces the SEALs to duck for cover behind the road's short perimeter barrier, along the gravel road.

Inside the White House Situation Room, it is 12:11AM Eastern Time, 07:11AM Baghdad Time.

The three video screens, each of which displays 40 independent video streams from each man in the three different SEAL Troops, is now a center of interest for the men and women in the Situation Room.

General Shane Maxwell is looking at Troop Charlie's monitor, where Colt Jackson is positioned. He points at the screen, "Oh, no! Look! Five of the videos just went out. ISIL must have sharpshooters. Headshots have destroyed the video feeds! Look, five of the streams have just turned black!
Lori Sanders feels like a knife went through her heart, "Huhhhh! Oh no."

Admiral Steven Whistler looks at the video wall for Troop Bravo---it is also split into 40 different streams. Six of its videos appear to be streaming the same image. Some of the streams are looking at a barrier and some are pointed at the sky. All of them show no movement, whatsoever. The Admiral thinks, "*Those poor men. They're going to be slaughtered.*"

On the west-side security perimeter in the middle of Haditha Dam, Colt assesses the situation. It is becoming grimmer by the minute. His squads report back casualties. In only the first volley, he has lost nearly 10% of his men. He thinks, "*We'll be decimated once the Caliph brings his Jihadists on top of the dam. It is impossible to stop them from getting on top---the dam is just too long and the access is just an easy walk up a gentle slope.*"

Colt Jackson, who had been looking over the south side of the dam before the ISIL volley of shots, COMMs Speedy Rabbit, "Where are you and your men?"

Lieutenant Michael Perry responds, "We're at your 9 o'clock, sir. We made it."

Colt Jackson speaks to the entire SEAL Team. "OK. Everyone, keep your heads down. Our best bet is to use the grenade launchers and attack the crowd below. Ultimately, they'll be forced to come up to the road, and then they'll have to funnel in. We can set up kill zones and use the structures in the middle of the dam for at least some cover. They will be completely exposed. Men, this is a fight to the death. Dig in, if we start beating them up, they'll launch their mortars and they won't care about weakening the dam. So, be ready to jump in your fox holes!"

SEAL Team 3 responds as if they were one body---with every man connected to the other, just like Siamese twins.

"HOOYAH, SIR...WE'RE WITH YOU!"

Colt Jackson screams from the top of his lungs,

"SEAL TEAM 3, FIRE!! FIRE!!! FIRE!!!!"

At this moment, one-hundred-nine rocket-propelled grenades are launched. They are strategically spread out among the thousands of Muslim fanatics, below. The grenades rip through the crowds, killing hundreds of fighters. Colt takes a look over the barrier and sees a mad dash for cover, below. He can see wounded men and can hear screams of pain coming from the Jihadists, below.

Colt Jackson orders"

"RELOAD!! FIRE! FIRE AT WILL! FIRE AT WILL!"

The SEALs fire off another round of grenades, but this time, the attack is less effective. The Jihadists have spread out quickly, making the grenade fusillade less effective. Colt takes another quick peak. A sniper's bullet rips concrete into his face. Colt holds his face and says, "Ahhhhh!"

At the White House it is now 12:20 AM Eastern Time, 07:20AM Baghdad

Time on the 28th of October.

Lori Sanders is looking intently at Commander Colt Jackson's video cam screen. She sees the bodies of dead Jihadists below, as a result of the grenades. She gasps as the image of the exploding wall in front of Colt Jackson's face---she witnesses the concrete blast directly into Colt's eyes. She cries out, "Huhhhh! Oh, my God!"

She and everyone else in the room hold their breath. A minute passes.

On the west-side security perimeter in the middle of Haditha Dam, Colt Jackson unconsciously calls out to anyone and no one, "Ughhh! Damn, that was close. The bullet hit the barrier! Thank God for this Q-Warrior head gear. I'm going to kiss the people who invented it."

Inside the White House Situation Room, Lori Sanders looks upward and says, "Thank you, Lord Jesus!"

Admiral Steven Whistler agrees, "Thank God."

The President also agrees, "Amen to that. Come on Colt, hang in there."

General Shane Maxwell offers verbal, but resigned support, "Be tough, Colt. Take as many of them with you, as you can." The General turns to the room and speaks to the group. "The poor sons of bitches. They'll be picked off in no time at this rate. The Jihadists haven't even deployed on the east and west sides of the dam, yet. All they have to do is crawl up the berm and come out on the road. They can stream in from as little as a quarter-mile away from Colt's defense perimeter."

Admiral Steven Whistler states the more than obvious, "Colt will be surrounded. They'll be shot at from the east, west and south."

Lori Sanders asks an important question, at least in her mind, "Can't they just swim away and come back another day?"

The Admiral responds, "Colt won't do that. He will try, at all costs, to keep control of that dam and break the back of the ISIL forces. Even if he has only a dozen men, he'll fight to the death. He won't abandon his men, or his mission."

Lori Sanders starts to cry. She turns to the President. "I'm sorry sir…"

and then to the Admiral and General, “I'm sorry General, sorry Admiral, it's just that we've been through so much with this man. This is how he's going to die?”

On the west-side security perimeter in the middle of Haditha Dam it is now 07:30AM Baghdad Time and 12:30AM Eastern Time on October 28th.

Colt sees more trouble heading their way. Black-robed figures are starting to come onto the road from both the west and east sides of the dam. Adding to his trouble, the Jihadists are driving American Humvees, which are completely bullet proof, straight up the berm of the dam and onto the road on top. Thousands of Jihadist fighters mass behind the vehicles. The clever Caliph has ordered two Humvees to the west and two Humvees to the east. They are slowly driving, side by side, toward SEAL Team 3. They are creating a shield for the ISIL forces.

Colt and his men are now becoming completely surrounded and Colt's kill zone defense has been rendered harmless by the Humvees. The bullets from the SEALs' FN MK20 sniper rifles and Colt M4A1 carbines, simply bounce off the reinforced armor on the Humvees.

Lori Sanders observes, “They've driven Humvees on top of the dam and the Jihadists are using them for cover! Those poor men, Colt will be slaughtered.”

The Situation Room is filled with grim faces. There is nothing they can do, but watch.

On the west side of Haditha Dam, behind the Humvees, Caliph Abu-Bakr is speaking to every cell leader in the now 19,000-man throng of Islamic Jihad fighters. “I will give the man who brings me Commander Colt Jackson alive, a dozen virgins---not in the next life, but in this one! I will give the man who brings me his scalp and his severed head, two dozen virgins. To the man able to capture him, scalp him and then behead him in front of me, I will give him six gold bars and 18 virgins!”

A war cry goes up from 19,000 raving fanatics, as word is spread from their cell leaders, to every single fighter. They begin to chant, 19,000-men strong:

أكـــبر الله! أكـــبر الله!

ALLAHU AKBAR! ALLAHU AKBAR!

Then in a frantic and demonic chant in support of their leader, the great Caliph Abu-Bakr, they shout:

العظيـــم نبينـــــا تحيــــا
بــو بكـــر أ الخليفـــــة!

Kaheeyah Nableeanah A'Valveem
Belfadeefah Ahdo Bakerra!

Long live our great prophet
Caliph Abu-Bakr!!

Inside the White House Situation Room, Lori Sanders gasps in complete and utter horror.

NSA's prism operations have always been able to pick up and retransmit local radio signals, from wherever the NSA has placed its satellite antennae. For this mission, nearly all eyes and ears have been trained on Haditha.

Lori Sanders, cries out, "OH MY GOD!! Mr. President!" She has uncontrollable tears flowing down her face.

Admiral Whistler talks to the screen, "Come on, Colt. Hang tough. Think."

General Maxwell answers, "I think we need to pray for a miracle."

At this moment, Lori Sanders defies government protocol and prays, out loud, in front of everyone.

"PLEASE DEAR GOD, PLEASE LORD JESUS, MAKE A MIRACLE HAPPEN. PLEASE HELP COLT JACKSON AND HIS NAVY SEALS AGAINST THE RAGING AND INFERNAL INFIDELS CALLED ISIL! IN JESUS NAME, AMEN!"

Everyone around the table, to Lori's surprise, responds, "AMEN!"

The President even pleas with the man upstairs, "We need a miracle now."

On the east side of the security perimeter in the middle of Haditha Dam, opposite Commander Colt Jackson and Charlie Troop, Lieutenant Michael Perry is facing the same threat. The Lieutenant estimates, about five thousand of the Caliph's men have been deployed to the east, ready to attack.

Lieutenant Michael Perry COMMs his Commander, "Papa Bear, this is Speedy Rabbit. I estimate a force of about five thousand approaching from the east. They are spitting distance from us, sir. They are amassed behind Humvees."

On the west-side security perimeter in the middle of Haditha Dam, Colt Jackson responds, "Roger that, Speedy Rabbit. On my mark, we're going to fire the grenade launchers and open up concentrated fire on them. Send two of your men and set up an IED. Rig the Improvised Explosive Device with some of the C4 recovered from the stockpile inside the dam. We'll do the same thing here."

Lieutenant Michael Perry replies, "HOOYAH, Sir. We're on it."

Colt Jackson mikes Lieutenant Cody Custer, "Tiger Fox, you and Stingray Charlie grab the C4 and wire it to blow in a shaped charge when the Humvees pass. Go, now."

Tiger Fox replies, "Sir, yes sir!"

Commander Jackson orders, "SEAL Team 3...OPEN FIRE!"

SEAL Team 3 opens fire on the Humvees approaching. The grenades detonate and slow the assault. Some of the grenades pass by the Humvees and streak into the growing mob of Jihadist fighters, hiding behind the protective cover of the vehicles.

At 12:36AM Eastern Time, 07:36 Baghdad Time, Kristi Kruger and Brian Pierce watch through the video cam of Tiger Fox. He is racing toward the Humvees. The suppressive fire from the SEALs has stopped the Jihadists from firing, momentarily. About 100 yards away from the front line of the west-side perimeter security area, he and Petty Officer Duke Moran, kneel down and set the charge. They turn to sprint back.

Petty Officer Duke Moran shouts over the noise of battle, "Piece of cake sir."

Tiger Fox shouts back, "Move your ass, Stingray Charlie. I don't want to lose you, today!"

The men race forward and are ten yards from the safety of the barricade and defense perimeter of Troop Charlie's position. After the concentrated fusillade of grenades and fire power from Troop Charlie, the Jihadists have composed themselves and return fire. Suddenly, Stingray Charlie goes down.

Petty Officer Duke Moran calls out, "I'm hit. I'm hit! Both legs---I can't move."

Tiger Fox stops running and returns to help his fallen comrade. Colt Jackson runs toward both men to help. A huge volley of gunfire and grenades streak from the SEALs' position, to try and protect Colt Jackson, Cody Custer and Duke Moran.

Lieutenant Cody Custer says, "I'm here, Duke, we'll get you out. You'll be OK."

Petty Officer Duke Moran looks up at Tiger Fox, "Don't let them behead me, sir."

Inside the White House, Lori Sanders, once again, gasps in horror.

Admiral Whistler says, "Save him, Tiger Fox!"

At the NSA, Kristi Kruger and Brian Pierce are riveted to the screens. Kristi unknowingly repeats the words just uttered miles away by Admiral Whistler, "Save him, Tiger Fox!"

Brian Pierce calls out, "Come on, Tiger Fox!"

Tiger Fox puts his arm around Stingray Charlie and they hobble back toward the protection of the SEALs' defensive perimeter. Colt Jackson races up and grabs the other side of Stingray Charlie.

Colt comforts Petty Officer Duke Moran, "Gotcha, Stingray Charlie."

They hobble as fast as they can back to protection. Bullets and explosions are filling the air like an overdone fireworks display. They dive for cover.

In the Situation Room, the group watches as blood pumps out of the leg of Stingray Charlie, in rhythm with the beat of his heart. Collectively, the room is horrified, even the battle-tested veterans present around the table, are grim with sadness. A major blood trail can be clearly seen from several of the SEALs' video cams.

Papa Bear grabs Lieutenant Cody Custer, "Tiger Fox, I order you to take care of Stingray Charlie. This Troop's medic is dead. Stay out of sight until you can get his bleeding under control."

"Aye, aye, sir."

Petty Officer Duke Moran looks up at Lieutenant Custer, smiles and assures him, "I'll be fine sir. It's only a scratch."

Then Petty Officer Duke Moran, AKA Stingray Charlie, closes his eyes and goes to sleep. The bullets have ripped through his femoral artery in his leg. With five liters of blood pumped through the heart every minute, Stingray Charlie has pumped his blood out all over the dam road.

On live camera, Stingray Charlie bleeds to death.

Kristi Kruger, observing the scene from the live video feed, feels her stomach getting nauseous. She turns to Brian and says, "I think I'm going to be sick."

Brian is feeling pain and sympathy for the men dying, trying to protect freedom all over the earth, he says, humbly, "That poor guy. What bravery."

On the west-side security perimeter in the middle of Haditha Dam it is now 07:46AM, Bagdad Time and 12:46AM Eastern Time on the most deadly day in U.S. Navy SEAL history.

Colt Jackson can now see the walls closing in on him. The sandstorm is still bad and has created a very strange sight. The sky is dark, but shades of light stream through the thick, black clouds enveloping the SEALs. The wind, the light, the sand, the dirt and the soot have created ghostly images on the battlefield in front of him.

The only good news is, the grenades have stopped the Humvees.

On the far west side of Haditha Dam, behind the Humvees, Caliph Abu-Bakr screams at the top of his lungs:

دفـعات عرب الهمفـي إلـى الأمام!

Lehfahra arabrut elhemtvee ila'enmahatn!

PUSH THE HUMVEES FORWARD! PUSH!

Colt Jackson sees the Humvees have stopped moving on his side of the Dam. He gets on his mike to Lieutenant Michael Perry, "Speedy Rabbit, were your men successful? Did the grenades work?"

"Yes, sir. The Humvees stopped, but sir, I think they're moving again."

Colt Jackson replies, "Let me take a look." Colt looks out on the scene, filtered through his infrared goggles. "My side's moving too. They must be pushing them. They are determined bastards, I'll give them that much."

Lieutenant Michael Perry answers, "Good thing we got the C4 in place, the shaped charges will keep the blast away from dam. That should take care of it."

Commander Jackson COMMs the order, "When they get to the C4, blow it!"

Two minutes later, two explosions come out from the surround sound of the video wall. Dust and debris fill the screens where video cams are trained on the ISIL Humvee positions. It takes a minute for the dust to settle, but the wreckage is complete. These vehicles will not be coming forward any longer and have completely blocked the passageway for any other vehicles. From here, at least the SEALs will finally clear a kill zone.

Colt Jackson COMMs his men, "Good work, Tiger Fox, Good work, Speedy Rabbit. Gentlemen, open fire, fire at will. They'll have to charge us now."

Inside the White House Situation Room Lori Sanders points to the screen, "Oh dear God! Look!"

Everyone looks at Colt's web cam screen.

No less than three thousand, black-robed screaming fanatics have leaped over the Humvees on the west side of the perimeter position. The road on top of the dam is narrow. This causes a traffic jam of human bodies and at least a thousand more black-robed warriors follow from behind. The Jihadists' AK-47s are firing streams of gunfire directly into the SEALs' positions. Several more web cam views are extinguished, and their screens in the Situation Room go black, the deadly result of a direct head shot.

General Shane Maxwell is beyond upset, "They're getting slaughtered. Look at all three Troop video walls. Half of the entire SEAL Team is down. Colt's lost a third of his men in less than an hour! DAMN IT!"

At 07:51AM Baghdad Time and 12:51AM, Eastern Time, Colt Jackson and SEAL Team 3 continue their defensive and deadly firefight against a total of ten thousand ISIL fighters who have poured onto the bridge. Five thousand poured in on the east side and five thousand on the west, with the SEALs killing at least a couple thousand of them, so far.

As Commander Jackson observes the battlefield, he sees exactly how the Caliph has split his force. Colt speaks to Tiger Fox, "The Caliph is smart. He's done what I would have done. He deployed from his original 20,000 man army, 5,000 on one side and 5,000 on the other, and then 10,000 to sneak right up on top of us from the south, over the top of the dam. He's using the cover of the storm to do what the Indians did to Custer, all right."
Tiger Fox looks at his Commander and says, "We'll die with our boots on and fight to the end, sir!"

"Tiger Fox, you know how much respect I have for you. I'll see you on the other side, regardless. Good luck, let's use might for right and take as many of them with us, as we can!"

Lieutenant Cody Custer tips his head to his Commanding Officer and friend, "It's been an honor serving with you, sir."

Colt replies, "Go get 'um, Tiger!"

Tiger Fox, "Roger that, sir."

Each man turns to the enemy and continues to pour 5.56 mm rounds, from their Colt M4A1 Carbines, into the sea of black robes surging toward them.

At NSA, Kristi Kruger bursts out into tears. Brian goes over to comfort her.

Inside the White House Situation Room, Lori Sanders breaks down, completely. She's sobbing.

The President has tears rolling down his face.

Admiral Steven Whistler has used his handkerchief to stanch his tears from flowing. His sadness overwhelms him and his effort to halt the tears fails. Streams of salty water roll down his cheeks.

General Shane Maxwell simply turns to everyone and says, "I can't look. These magnificent men are all going to die. God love them."

Back on the west-side security perimeter in the middle of Haditha Dam, Colt reads the clock on his Q-Warrior head gear. It is 07:59AM Baghdad Time and 12:59AM Eastern Time.

It has only been 56 minutes since the first shots of this battle took place. Colt feels like it was ages ago. Suddenly, a blood curdling cry comes out from the 5,000 Jihadists, whose front line is now only twenty yards away from Colt Jackson and the SEALs on the west. Another 5,000 also scream out a war cry. They are only twenty yards away from Lieutenant Michael Perry, to the east.

The wind has abated and the SEALs can literally see the whites of the enemy's eyes. Two thousand dead ISIL bodies have fallen on each side of the defensive perimeter, totaling over four thousand dead bad guys.

Now, with far less than 100 men left, the enemy is getting ready to overrun their positions. The stacks of bodies on the road in front of the SEALs slow, but do not stop, the Jihadists' assault.

Suddenly, to Colt's left and rear, the first Jihadists breach the barrier wall.

The ISIL fighters have climbed up the front side of the dam's gently sloping, earthen wall. SEAL Team 3 now also has to engage in mortal, hand-to-hand combat with the enemy. Every Jihadist wants their 18 virgins and a pot of gold, it seems.

Tiger Fox yells to Papa Bear, "They've breached the main barrier coming up from Haditha, below, sir. We're going to be hand-to-hand!"

Colt Jackson turns to Lieutenant Cody Custer and says, "Tiger Fox grab twenty men…" Colt points to a group of SEALs. "…Get up on the barrier and start mowing down the Jihadists climbing up the side of the dam! MOVE IT!"

Tiger Fox replies, "Sir, yes sir!"

Tiger Fox and a group of men open fire on the black-robed figures climbing over the top of the barrier. Tiger Fox and his twenty men reach the barrier and set up skirmish positions. They start raining lead down on the enemy below.

Master Chief Jack Flintlock, call sign Killer Whale, takes his position on the right side of Tiger Fox. He looks at the scene, below. "My God sir, look. There are thousands of them. It's just like that movie with Brendan Frazer. What was it? Oh yea, The Mummy. But, these guys won't crumble into dust, though, will they sir? Too bad, we'll just have to do it the old fashioned way!"

Tiger Fox smiles in the middle of the battle at his Master Chief and says, "Roger that, Chief! Keep your head low and keep on firing! We may as well get as many as we can. LOOK AT ALL OF THEM! They're like an army of ants. They just keep coming!!"

Inside the White House Situation Room, no one has ever witnessed a coming massacre like this, before. Even Admiral Steven Whistler is losing his composure.

Admiral Steven Whistler pleads with God, above, "Please, God. Help them."

At Kristi Kruger's and Brian Pierce's computer pod, Brian looks at Tiger Fox from the viewpoint of Killer Whale's camera. He stares at his utter bravery in the face of impossible odds, and the overwhelming force before him. Brian looks at Kristi and says, "Kristi, he's a better man than I am, or I'll ever be. Just look at them fight against all odds---and with no help. These men are magnificent!"

Kristi, with tears in her eyes, is inconsolable, "I feel so helpless." Tears pour down her cheeks. She keeps staring through the eye-of-the-lense of

Lieutenant Cody Custer's head cam. She watches with grim terror as the sea of black-robed murderers relentlessly press forward toward this group of brave U.S. Navy SEALs. She doesn't even realize she is speaking out loud. Shock is setting in, "God, please help them!"

Kristi is now sobbing uncontrollably and collapses in Brian's arms.

Inside the White House, every person in the room has their hands over their mouth. Everyone can taste their salty tears.

On the west-side security perimeter in the middle of Haditha Dam, Colt Jackson screams at the top of his lungs, to anybody and everybody with a rush of adrenaline:

"COME ON, YOU SONS OF BITCHES!
COME ON YOU PANSIES! COME AND GET US!!"

Colt and what's left of his western perimeter team, lays down a blistering and lethal round of cross fire from their Colt M4A1 carbines and FN MK20 sniper rifles, but now, some of the men are down to only their MK23 H&K handguns. The firefight has been the most intense anyone on SEAL Team 3 has ever experienced.

Dozens more Jihadists drop, only to be replaced with even more black-robed, wide-eyed fanatics.

All of them seek Allah's prize and the gift from the Caliph.

Colt speaks into his microphone, so the entirety of SEAL Team 3 can hear him. "Gentlemen, I'm sorry I can't even call for a 'Broken Arrow' to direct all fire power directly on our own positions. We have no help. I'm sorry, men."

SEAL Team 3 automatically rallies to their leader, "HOOYAH, SIR!! TO THE DEATH, WITH HONOR AND GLORY, SIR!"

Commander Jackson screams, "HOOYAH! TO THE DEATH, WITH HONOR AND GLORY!"

The only thing that can be heard from Kristi Kruger's and Brian Pierce's computer pod at the NSA, is the sound of blistering gunfire and hysterical crying---coming from both Kristi Kruger AND Brian Pierce.

At 08:01AM Baghdad Time and 01:01AM Eastern Time, Lori Sanders is an emotional train wreck. Tears stream down her cheeks and uncontrollable discharge is coming out her nose. Her stomach is torn apart inside. She can hardly breathe and she buries her face in the table of the Situation Room. She slams her fist against it and cries her eyes out.

Everyone else's face is filled with tears and cries of grief also. They watch as a dozen more web cam videos become extinguished on the video walls before them.

Back on the west-side security perimeter, Colt and his men continue to fire, but the oncoming surge from the Jihadists is just too great. When they fall and pass on to their virgins in heaven, the black-robed fighters are immediately replaced by other advancing fanatics.

Colt Jackson unconsciously speaks what he is thinking, "They're like sharks teeth. One breaks off and it's immediately replaced by another. Commander Jackson continues to urge his men on, regardless of the lost cause before them, he COMMs the entire Team:

"KEEP FIRING MEN. HOLD YOUR GROUND!!"

All the SEALs scream at the top of their lungs, "HOOYAH, SIR!" Colt can hear the echo of the same response from the east side of the perimeter from Speedy Rabbit's men. He thinks:

"God almighty, thank you for blessing me with these men.
I am so proud of them!"

Unfortunately, the SEALs are starting to run out of ammunition.

Tiger Fox pulls the trigger on his carbine and it just "clicks." Colt Jackson experiences the same problem.

"Papa Bear, I'm out of rifle ammo, switching to my handgun."

Colt replies, "Me too, Tiger Fox. Make every bullet count. This is going to be it. It has been an honor to serve with you."

The two men look at each other from their positions---Colt on the front line of the west side of the perimeter and Lieutenant Cody Custer on the south wall overlooking Haditha. They tip their helmets with respect to each other.

When Colt tilts his head he thinks he hears something toward Lake Qadisiyah to the north, bearing down on his right flank. He concentrates, he thinks:

"*I do hear something. It sounds like the roar of engines.*"

Commander Jackson COMMs Lieutenant Cody Custer, "Tiger Fox, can you see anything on the lake?"

Tiger Fox replies, "I'm not sure, maybe sir, but I *HEAR SOMETHING*!"

Through the fog of battle the sound is distinct. Colt hears the familiar sound of multiple .50 caliber machine guns. They are indeed coming from his right flank on Lake Qadisiyah.

Colt knows they are dead men, sooner rather than later, if the Caliph has now also surrounded them from the water. The water has kept SEAL Team 3 protected on at least one side. If the Caliph is now initiating an attack from this position, on their blind side, this battle will be over in minutes.

Colt Jackson COMMs the entire Team, "Brace yourselves, men. I think they've got us boxed in from the Lake. I heard a .50 caliber machine gun. I'm certain of it. Keep firing, but watch your right flank on the west end and on the east end of the perimeter, watch your left flank."

The sand made it difficult to see, but now the sounds Colt was hearing are closer. He realizes exactly what is in the water.

Commander Jackson COMMs the entire Team and shouts into the mike, "Courage men! The cavalry has just arrived! The SWICKs are here!! IF YOU'RE LISTENING---THANK YOU ADMIRAL!! KEEP FIRING MEN. The SWICKS will knock their guts out and give us a fighting chance!"

A very loud war cry comes up from the U.S. Navy SEALs trapped on top of Haditha Dam:

**"HOOYAH, SIR!! DEATH TO ISIL---
DEATH TO *THOSE* INFIDELS!!"**

Colt Jackson has accidentally left his COMM open. He speaks out loud, repeating his earlier thoughts to himself, "God, I'm proud of these men!" Commander Jackson, of course, is also broadcasting live back to Washington D.C. and Ft. Meade, Maryland.

U.S. Navy Special Warfare Combat Craft ("SWCC," nicknamed "SWICKS") are perhaps one of the best kept, open secrets, of the SEALs. These craft and their highly-skilled crewmen, engage mainly in clandestine infiltration and exfiltration of SEALs and other special operations forces. These SWICKS now are aiding the trapped SEALs with amazing firepower.

Their electrically powered, GAU-17/A miniguns allow bursts of up to 6,000 rounds a minute, potentially making a major dent on ISIL forces. If these guns pour out lead at this rate, on a continuous basis, their barrels would literally melt. When Commander Jackson heard the heavy .50 caliber machine gun fire and its distinctive "thunking" sound, he knew it only told half the story. The pressure released from each shot is powerful and is actually felt by the gunner and those around him. It feels like somebody has nearly knocked the wind out of you, if you're not used to it. Arguably, there is more directed firepower, per square foot, on a SWICK, than any other military vehicle.

Colt Jackson and his men continue to fight. There are four SWCC's in the reservoir. Two are on Colt's side, trying to fend off ISIL to the west, and two are on Lieutenant Michael Perry's side, trying to push them back to the east. The SWCC's are mowing down thousands of Islamic radicals. The bodies stack up on both sides of the SEAL perimeters.

Colt and his men thought this would have deterred the Jihadists, but somehow, they keep coming. Now, they are beginning to overrun the top of the barrier from the south, which Lieutenant Cody Custer is defending with his team. There are nearly nine thousand more fighters coming from this direction, aiming to breach the top wall of the dam.

Colt Jackson says to his men, "Hang tough, men. We have a little hope."

Inside the White House Situation Room it is 08:03AM Baghdad Time and 01:03AM Eastern Time. Lori Sanders hears the chatter and picks up her head off the table.

Lori says, "What is it, what's happening?"

Admiral Steven Whistler responds, "Looks like somehow, SWICKS arrived on the reservoir. They're trying to help Colt and the SEALs, by firing into the ISIL positions on the west and east flanks."

The Admiral picks up his combat phone and speaks loudly to his adjutant. "Find out how those SWICKS got to Lake Qadisiyah!" He hangs up the phone.

General Maxwell is taking in the battle, "It might buy them some time, but they are still outnumbered, by thousands-to-one odds."

On the west-side security perimeter, the wall of sand makes everything appear as a ghost-like image. Colt takes a look on the lake. He can clearly see two American SWCCs, with their crewmen blasting away at the enemy. Actual flashes of fire are coming out of the barrels of the guns, like Colt has never seen before. Colt reckons the SWICK crews have pushed it to the limit. Those barrels could melt down if they keep up this rate of fire.

Then an amazing thing happened.

Colt Jackson perks up. He shouts over the COMM to his men, "Men! Listen! Listen to what those boys are playing! They're playing the "Marine's Hymn." Some of the crewmen must have joined the SEALs from the Marines. They know that's where a lot of us came from! LISTEN!!"

From the SWCC, closest to Commander Colt Jackson, and also a Colonel in the United States Marines, the music blasts:

"FIRST TO FIGHT FOR RIGHT AND FREEDOM.
AND TO KEEP OUR HONOR, CLEAN. WE ARE PROUD TO
CLAIM THE TI-A-TAL, OF UNITED STATES MARINE!"

Every single living U.S. Navy SEAL on the Haditha Dam screams to the top of their lungs:

HOOYAH!!

The men's confidence soars. They continue the battle and keep on fighting.

Inside the White House, the Situation Room is stirred by what everyone hears from the speaker system integrated into the video wall. Everyone gets goose bumps when they hear the Marine Corps Hymn over the SWCC's loud speakers.

But, as the Situation Room feels the surge of patriotic pride and emotion, another half-a-dozen video cams go black on Troop Charlie's video wall.

The SEALs are still getting decimated.

On the west-side security perimeter, Colt knows it's only a matter of time. He looks over at Lieutenant Cody Custer. He has taken out his DOC Flipper knife and is about to kill a Jihadist who was about to fire on Master Chief Jack Flintlock.

Tiger Fox screams, "Not so fast, you bitch!" He plunges his knife into the insurgent's spine, killing him instantly.

Master Chief Jack Flintlock takes his eyes off the battle for a split second and looks at his Lieutenant, "Thanks Tiger Fox."

Lieutenant Cody Custer just winks at Killer Whale and keeps on fighting. More Jihadists keep pouring over the top of the barrier and are now intermixed with the SEALs, on top of the dam.

It is now 08:20AM, Baghdad Time, 01:20AM Eastern Time. The SWICKS have temporary halted the Jihadists' advance from the east and west, beating them back on both fronts. This allows Colt and his SEALs, who were pointing out toward the western front, to pivot south and sprint over to Tiger Fox, whose unit desperately needs help.

Colt and his band of brothers open fire at close range on the intermingled Jihadists. They all drop like stones with the SEALs' MK 23 .45 caliber handguns.

Colt eases up next to Tiger Fox. He looks down, over the wall of the barrier. Colt Jackson views the scene, and says, "Holy shit! There are thousands of them!"

Tiger Fox replies, "Yes sir. We have to stop firing when they pour over the top and kill them up here. They are whittling us down."

Colt Jackson and Tiger Fox survey the situation. They estimate there must be at least 7,000 men coming up the slope of the berm, of the earthen dam wall.

Tiger Fox looks grimly at Papa Bear, "I'm almost totally out of ammo, sir."

Colt Jackson is in the same situation, "Roger that. Me, too."

Inside the White House Situation Room, Lori realizes the SWICKS can't shoot through the dam to the south, where the invading forces of another 7,000 radicalized Jihadists are pressing their attack.

Lori Sanders exhales, resignedly, "Oh, NO!"

Colt Jackson and Tiger Fox pick up carbines from two of their fallen comrades. They continue to pump iron out of their newly acquired rifles, together. The other members of the "Heaven's Seven" are lined up next to each other. This includes their fallen comrade, Petty Officer Duke Moran, call sign Stingray Charlie.

The men just unconsciously move together, as a unit, and stand unified in one last stand. They trained this way and when the chips are down, they always come together, as a team. This morning was certainly no exception.

They look at each other---they are all filthy. The roar of gunfire, the soot of the battle, the dirt and sand from the black storm and the sound of battle cries, makes this a fight of the ages---captured completely on live TV.

Colt Jackson attempts to fire his Colt M4A1 Carbine, but he's out of bullets:

"Click...Click...Click."

Tiger Fox is continuing to fire, but when he continues to press the trigger, he also runs out of bullets:

"Click...Click...Click."

"I'm out Papa Bear."

The men look at each other. They, again, both draw their MK23 .45 caliber handguns. This is all they have left. They begin firing over the barrier, making each shot count. They are facing thousands and thousands of black-robed, evil killers.

At the NSA, Kristi Kruger is sobbing. "I'm sorry, Brian, I just can't stop crying."

Inside the White House Situation Room the President looks with admiration and pride at the brave men. "Look at them, brothers to the end. Unbelievable."

Lori Sanders is biting the index finger of her right hand and is crying, nearly uncontrollably.

While the group in the Situation Room is watching the final moments of the coming massacre, the surround sound system of the video wall starts to receive interference with the audio streams coming from the men in Charlie Troop. They can barely hear a word Colt Jackson is saying over his mike to his men.

Simultaneously, Colt Jackson also suddenly can't hear anything! Something extremely loud is preventing him from hearing his men---even those right around him.

Suddenly he feels the fierce blast of a gust of wind, traveling at hundreds of miles an hour. It knocks him into the barrier. Colt looks up into the ghostly sand and soot-lit sky of the black storm above.

There, Colt makes out the familiar shape of an Apache helicopter and Colt's COMM comes to life.

"Papa Bear, Papa Bear, this is Scepter 2, over."

From Tiger Fox's video cam, both the White House and the NSA can see the picture of a tear running down the face of Commander Colt Jackson.

He knows it is his pal, Prince Harry! Prince Harry is hailing Commander Colt Jackson over his mike and SEAL COMM line. Somehow he got here to save his ass and what remains of SEAL Team 3.

He squawks up to his friend, "Scepter 2, you are a site for sore eyes.

Great to see you mate."

Prince Harry replies, "Great to see you, too, mate. I've brought along a few of my friends."

At this precise moment, a FLEET of Apache Helicopter Gunships flies over the dam. Colt counts at least half-a-dozen choppers, maybe more.

The White House Situation Room erupts in applause and shouts of joy.

The President screams at the top of his lungs, "WAY TO GO, HARRY!!"

Admiral Steven Whistler adds, "That son of a bitch. I don't know how he did it, but I'm going to kiss that guy, someday!"

Lori Sanders is thanking the good Lord, upstairs, "THANK YOU LORD JESUS!"

General Shane Maxwell then says, "and thank Prince Harry and his boys, TOO!"

Lori Sanders responds with joy, "Believe you me, sir, I WILL!" She then pleads with the video wall, "Please get them out of this mess, Scepter 2!"

At Kristi Kruger's and Brian Pierce's computer pod, Kristi says, "I didn't realize Commander Jackson knew Prince Harry. Did you?"

Brian answers, "No way. That one's above my pay grade."

Colt yells into his microphone, "Go get 'em Scepter 2!"

Prince Harry replies, "We thought you could use a little help. You didn't think I was just going to sit around and let you have all the fun, today, did you?"

Colt Jackson starts laughing, the strain of battle is relieved by humor---and Prince Harry knows it.

Commander Colt Jackson then replies with the true bravado of a U.S. Navy SEAL and the exact same bravado Prince Harry demonstrated when he was wounded and trapped in his Apache in Afghanistan.

Colt responds, "We've got 'em under control---right where we want 'em. We've got them surrounded. They're easy to shoot at this range, especially when we can see the whites of their eyes!"

Prince Harry laughs a quick laugh and says, "Yes, mate. It sure looks like it's all under control! I'll be right back. Me and the boys have some grass to mow. Cheers mate. Tell your men to keep their heads down and have them focus their fire power on the top of the dam. We'll take care of the rest."

Colt Jackson, now serious, gives Scepter 2 an important piece of information, "We're down to our pistols, Scepter 2…and thanks, pal."

Prince Harry responds, "Roger that. Cheerio...won't be but a tad!"

Inside the White House Situation Room, Lori Sanders cries out, "That man will be the life of me, I swear. How can they be that calm and laugh off near-certain death like that? UNBELIEVABLE!"

The President announces, "I'm calling the Queen and thanking her, immediately."

The President gets on the hotline to the Queen and is on the phone with her, quickly thanking her for Prince Harry's brave actions in saving SEAL Team 3.

Inside Buckingham Palace in the Queen's bedroom it is 06:27AM London Time, 01:27AM Eastern Time and 08:27AM in Baghdad.

"Hello, Mr. President, what can I do for you this early morning?"

The President answers, "I'm sorry for the early morning phone call, your highness. But I wanted to thank you for your support. Thanks for sending in Prince Harry and his mates in their Apaches to bail out SEAL Team 3."

The Queen of England is now more wide awake and then asks the President, "Where did you say they were?"

The President looks around the Situation Room at the live scenes from Iraq and says, "They're in Haditha, in Iraq. There's quite a firefight going on and Harry is saving the day!"

The Queen stoically states, "Thank you, Mr. President, but I'm afraid that's *quite* impossible. Harry is on vacation with his mates. They all took off for Switzerland to do some fall hiking. I think they left a couple of days ago."

The President responds, "He did, huh? Hmmm. Maybe I'm mistaken. Sorry to wake you up and sorry to give you a scare."

The Queen answers, "Mr. President, when you get to my age, you don't frighten that easy. You have a good day."

"You too, your highness."

Admiral Whistler inquires, "What did the Queen say?"

The President fills them all in on the phone call, "She said Prince Harry was in Switzerland. He left a couple of days ago."

General Maxwell raises his eyebrows, "Are you serious? How in the hell did Prince Harry get to Haditha, then?"

Everyone looks around the room.

Lori Sanders says, "I don't know, but let's hope he can help Colt."

All eyes return attention to the battle before them.

Inside the Apache helicopter gunship, Prince Harry leads his team of men, "OK, mates. Let's mow them down, split the squad in two. We'll wipe out the right and left flanks of the ants crawling up the wall of the dam, then one team will go west and one east. We'll finish the job the SEALs and the SWICKS started. Let's route these bastards---once and for all!"

Harry's Apache helicopter mates, with distinct, calm and cool British accents reply, "Let's do it, Scepter 2!"

At 08:40AM Baghdad Time and 01:40AM Eastern Time, Colt and his team look down as the Apaches literally mow down their foes. The 7,000-man-strong ISIL army, crawling up the south side of the dam wall, is crumbling before their eyes.

Colt Jackson gathers his strength, his men, and then he orders, "WE'RE

GOING TO ATTACK! I repeat ATTACK. Speedy Rabbit, attack east. We're heading west. Let's break their backs on my count. 3...2...1...CHARGE!"

Colt and his men charge into the western front of the ISIL warriors. The SWICKS continue to blast away at the black-robed fanatics. The Jihadists' begin to retreat. Colt sees one man barking orders. He recognizes Caliph Abu-Bakr from the Marsa Malaz at The Pearl, in D'oha, Qatar.

Colt's heart races.

Commander Colt Jackson is out of bullets and he races toward, and catches the Caliph. He tackles him from behind.

"You piece of shit, take that!"

Colt rips his knife from its sheath and plunges the Microtech DOC Flipper dagger deep into the heart of Caliph Abu-Bakr. Colt looks around fiercely at several ISIS warriors, with his bloody knife in his hands. He waives it at them fiercely, like the brave warrior he is.

The Jihadists raise their weapons, getting ready to fire at Colt.

The Islamic Jihadist Fighters all retort, "ALLAHU AKBAR!! ALLAHU AKBAR!! Take this you Yankee piece of shit!!"

They take aim, getting ready to execute Colt Jackson.

Inside the White House Lori Sanders is incredulous. She cries bitterly, "He could have just stayed safe! Oh my God! NO! NOT AFTER ALL THIS!"

Admiral Whistler tries to console her and explains, "Ms. Sanders, he's not paid to stay safe. He never quits. He never has and he never will. God bless Colt Jackson."

At 08:42AM Baghdad Time and 01:42AM Eastern Time, Colt hears the whining sound of the spinning machine guns from a very angry Apache Helicopter. The six Jihadists, who are pointing their weapons at Colt and are lined up in front of him, have their heads blown off.

A perfect hit, just in the nick of time, by Colt's old pal, Scepter 2.

Prince Harry squawks to Colt over the open COMM line, "I guess they lost their heads over that one, hey, mate?"

Colt Jackson stares directly into the cockpit of the hovering Apache, which is no more than twenty yards from Colt's position. He tells his friend, "You are one hell of a guy, Scepter 2. One hell of a guy. I'll have to buy you a beer later on, today."

Apache Helicopter in Sandstorm

Public domain image. Work product of U.S. Government. Artistic affect by Robert L. Kelly.

Prince Harry responds, "You're on! Let's clean up this mess. Looks like they're finished. Say, mate, one of the friends of our family is having a get together in Switzerland in the next couple of days. Why don't you come along? I think you know her. You do remember Mary Barrister, don't you?"

Colt replies, "Of course I know Mary, Scepter 2. That sounds like a GREAT idea. I haven't seen Mary in AGES!"

In the Situation Room, Lori Sanders, all of a sudden, feels a completely different set of emotions. She is crushed! After all this, Colt was going to run into Mary Barrister's arms. She starts crying, again.

Lori begins to cry, again. Emotionally, she is drained beyond what most people are capable of handling. She quizzically says quietly to herself, "Switzerland! *What*?"

The President is victorious, "Is everything all right, Ms. Sanders? You don't need to cry, although everyone here today, has shed a tear. There is no shame in that. Thank God, Colt and the rest of the Team will be all right! Admiral, you've won the day. You've broken the back of ISIL!

Congratulations to you and the General!"

Admiral Steven Whistler responds "I think the congratulations should go directly to those young men up there." He points to the video display screens. Half of SEAL Team 3's video cameras are blacked out. He continues, "This was a mighty victory, but at a terrible cost."

General Maxwell responds, "I've never seen such bravery in the face of such overwhelming forces. Admiral? How did those SWICKS and the Apaches get there?"

The Admiral answers, "I honestly have no idea. I don't know."

General Maxwell instructs Ms. Sanders, "I want you to pull up video satellite surveillance for the area. How did those guys get there? And how come nobody knew about it??"

The President now understands what is going on, "Wait a minute. You guys didn't order the SWICKS, or the Apaches into battle?"

The Admiral and the General reply, "No, sir."

The President asks, "Well then, who in the hell did?"

CHAPTER 25

FINAL CONQUESTS

The next day, October 29 it is 04:00PM in Ft. Meade, Maryland, home of the NSA's headquarters. Kristi Kruger and Brian Pierce are busy cleaning up the computer pod after the intensity of the last day-and-a-half.

Kristi is still shaky from the events of battle. She asks Brian, "Those men really are special, aren't they?"

"Yes, Kristi. They are---every one of them." Brian Pierce sighs and speaks in a defeated tone of voice, proud of the bravery the SEALs displayed, but discouraged because he knows he has no chance with Kristi. He continues, "I think I'm just going to leave early. I'll see you on Monday, OK?"

As Brian leaves, Kristi gets his attention, one more time, "Brian?"

Brian turns around from the entrance of the computer pod to face her, "Yes?"

Kristi looks affectionately at Brian and says, "Thanks for being there for me. During the last two days, I realized many things. I'm not sure I could handle it if I was with someone who had to take those risks the SEALs do every day of their lives."

Brian Pierce still feels defeated, as Kristi continues to talk about the SEALs and how brave they are.

Brian, as gently as he can, cuts her off and says, "No problem, Kristi. I'll see you Monday, all right? I don't think Lori will mind if we each take tomorrow off---it is Sunday, after all."

Brian, again, turns to leave.

Kristi, one last time, gets Brian's attention, "Brian? Remember I told you when this was over I had a question to ask you?"

Brian looks at the gorgeous face and beautiful figure of Kristi Kruger. He answers in a kind voice, knowing her delicate emotional state after witnessing the harrowing experience of the SEALs yesterday, "Yes? What is it, Kristi?"

Kristi looks at him with her deep, chocolate-brown eyes. She smiles and asks, "Will you have dinner with me?"

Brian Pierce could have clicked his heals in the air like a Leprechaun. He grins from ear to ear and answers, "Boy would I! Hot, diggity dog!"

Kristi Kruger just smiles, laughs, and shakes her head at some of the strange expressions he uses, sometimes. She says, "Great, how about tonight? It's Saturday and it's only 4 o'clock. Can you pick me up at 8?"

A delighted Brian Pierce replies, "YOU BET!" He is now among the happiest men on earth and he floats out of the NSA. He thinks,

"*Finally, I get to have dinner with Kristi Kruger*!"

Two days later on Halloween, Lori Sanders is walking on N Street in Georgetown, where she owns a townhouse. She walks alone on her street filled with sadness.

She has been so distraught over Colt Jackson, she took the day off from work. She had hoped a three-day weekend might help. Emotionally, she was exhausted. She was also heartbroken.

Word had come back through the Admiral that Colt Jackson had been given a three-week pass and he headed off to Switzerland with his pal, Prince Harry.

Lori was trying to set her mind on other things and was returning from some of her favorite local shops on Wisconsin Avenue in Georgetown. Her head is bowed down and she has tears in her eyes. She sniffles in the brisk air of the Halloween evening. She approaches her townhouse on tree-lined N Street, just down from Paulo's.

Lori is crying. Her eyes are swollen and her vision is somewhat blurry. Before she sees reaches the big cherry tree in front of her home, she sees some little kids pass her in goblin suits for Halloween. They notice she is upset.

A little boy holds up a Snicker's bar for her.

He says, "Here you go, lady. Don't cry."

Lori sniffles and smiles. She accepts the kind gift from the little boy. "Thank you very much. You're very sweet."

The little boy walks off and turns his head around, looking at the sorrowful figure of Lori Sanders. He says, "Don't worry, lady, everything will be all right. The Snicker's bar will cheer you up! Bye, bye, now---I have to get more candy."

Lori watches the sweet little boy cross the street with his group, "Bye, bye. Thank you and be safe."

The little boy crosses the street, with his group of friends and their chaperones. Lori arrives at her townhouse and sees a dark-clad figure sitting on her front, red-brick door step. The person is sitting directly under her cherry tree. She wishes it was blossoming now, because it is beautiful when it blooms in the spring.

The figure gently calls out, "Don't cry, lady. Everything will be all right!"

Lori Sanders is shocked and she cries out, "COLT! What happened? Where did you go? You disappeared. We were all so worried! What happened after the battle?"

Colt replies, "When the battle ended, the entire SEAL team took care of our wounded, and made sure our dead brothers were flown back home. It took us a couple of days to do just that. Prince Harry and his mates helped us."

Lori Sanders says, "They did?"

Colt looks at Lori, "Yeah, they did. In fact, I had a chance to have a long chat with Prince Harry. He told me he received an urgent phone call from a mutual friend of ours…you do remember Commander James Forrestal, don't you?"

Lori Sanders looks around, somewhat nervously, "Huh...yes, I do."

Colt continues, "The strangest thing happened to the Prince."

Lori asks, "What was that?"

Colt goes on, "I guess James told him to expect a call from an anonymous, but top-secret source, with an urgent message. James told him it would be coming from a woman, but to protect her, no one could know her name---not even me. James apparently received the phone call from this mystery woman, a couple of days before the battle. I think she threatened him with his career if he ever divulged her true identity."

Lori Sanders looks up in surprise, "REALLY?"

Colt Jackson looks deeply in Lori's eyes, fixing his cobalt-blue steel gaze upon her, "Yes."

Lori asks, "How did Harry even get there through the black storm?"

Colt answers, "It turns out, Harry was in Switzerland vacationing. But, then he received the phone call. Apparently, the information provided to him was mission specific and clearly authentic. Harry immediately went into action, to save his pal. That pal was me."

Lori Sanders feigns surprise, "REALLY?"

Colt continues, "That Harry is really something. He got on the horn with his boys in the 656 Squadron Army Air Corps, from England. It is the most prestigious aviation unit in their country. He told them what was going on and said he wanted to make sure I had back up, in case the plan to destroy ISIL didn't work."

Lori Sanders tries to act incredulous, "You've got to be kidding me? HE DID?"

Colt Jackson continues to stare, deeply, at Lori. "Yes, he did."

Lori then asks, "But, how did he get Apaches through the sandstorm and how did those SWICK's get on Qadisiyah Lake?"

Colt explains, "Harry and his men commandeered the Apaches from the forward air base in Mosul, where both the Americans and the British have been supporting the Kurds."

Lori asks, "What do you mean, commandeered?"

Colt answers, "They stole them. Well, let's just say they borrowed them."

Lori continues her quest for knowledge, staring directly into Colt's cobalt-blue eyes, with her sparkling, emerald-green ones, "But, how did he fly them through the sand?"

Colt responds, "It turns out, he didn't. Apparently, as the storm was building up, before it hit Iraq, this same woman called him on the phone and told him about the weather coming in from Turkey. She told him there could be a hellacious sandstorm. Harry had the Apaches loaded on trucks, which he also had commandeered, and they drove them down from Mosul. It took them just under three hours and they had an armed escort all the way."

Lori, the intelligence agent needs to know more, "But they still flew in the black storm. *How did they do that?*"

Colt puts the pieces of the jigsaw puzzle in place for her, "The mechanics rigged the engines with heavy-duty sand filters. Harry knew they would only give him and his men a limited amount of flight time during a major sandstorm, but he took the risk of crashing and burning to save my life and the lives of over sixty other men. He and the boys from the 656 are heroes."

Lori answers, "They are brave and they are heroes. Colt, we saw it all…I was…I was…I thought you were going to die." A tear runs down her face. Lori takes a deep breath and presses for one more answer, "But, what about the SWICKS? Where did they come from?"

Colt stares deeply at her. He admires her deep need to always know everything, because her brain just worked this way. Colt knows she is brilliant…and lovely---on the inside and out.

Colt answers, "It turns out our pal, Commander Forrestal, further risked his career and also took a great chance for us. He called Vice Admiral John Stiller."

Lori shoots him somewhat of a dirty look and says, "By the way, I know about your flirting with Akiko!"

She smiles at him.

Colt smiles back, mischievously, and says, "Well, I had to learn to be an actor on that one. We're going to leave her on a string, by the way. She might be able to be turned, if we handle her the right way. Anyway, when the Admiral heard what was happening, he commanded four Chinook Helicopters to ferry four SWICK's up to Lake Qadisiyah. They arrived the morning of the 28th, just before the sandstorm hit, but they were dropped into the water in one of the jagged coves on the far side of the reservoir, because the Admiral didn't want to alert the enemy to their presence. The SWICK's arrived as fast as they could--just in the nick of time."

Chinook Helicopter Carrying a SWICK

Public domain image. Work product of U.S. Government.
Artistic affect: Robert L. Kelly

Lori tells Colt, "Everyone was filled with pride when the SWICK crew blasted out the Marine Corps Hymn. It gave us goose bumps."

Colt responds, "It was like hearing a cavalry charge in the old West, when surrounded by Indians, and the wagon train was brought into a circle for protection. This is exactly how we felt. We got super-human strength from that song. There is no doubt about it."

Lori speaks of the SWICK boat heroes, "Those are some really great men, Colt. Thank God they were there to help you and SEAL Team 3. We were all so worried. She then asks him nervously, "Are you going to Switzerland?" She knows Colt realizes she heard every word Prince Harry said to him about Mary Barrister and vacationing in Switzerland, together.

Colt continues looking deeply into Lori's beautiful, bright-green eyes. He takes in her light-brown, blond-streaked hair and then looks down---afraid to tell her he is going to see Mary Barrister, "Well...yes, I am, Lori. I will leave right after the funerals. In fact, the Admiral has given me a furlough for a few weeks and I thought it might be really nice to see Switzerland, hang out with Harry for a while and yes, see Mary."

Lori looks down dejectedly and tears, again, come down her face. She sniffles and says haltingly, "I...U-N-D-E-R-S-T-A-N-D...I'm so sorry I called you thick-headed and treated you that way during the meeting..."

Colt Jackson looks at her. He doesn't blink an eye and holds her in his stare, "I have to ask you a question."

Lori sniffles her nose again, and weakly asks, "What is it?" Tears are running down both her cheeks and her mascara is running. She feels so weak.

Colt simply asks her, "Why did you do it? Why did you risk everything for me and the Team?"

Lori now looks, with her glazed-green eyes, deeply into his. She knows Colt has figured out she was the one behind Prince Harry, and even set the ball in motion with Commander Forrestal. She sighs, and despite her defeat on the romantic front to Mary Barrister, she takes a deep breath of courage. She says,

"Because I love you and I didn't want to see you DIE!"

Her eyes widen emphatically. She starts crying again.

Colt Jackson takes both his hands and gently takes hold of her face. He strokes the tears away and Lori's mascara smears all over her cheeks. Tears continue to flow from her.

Colt gently answers, "That's what I figured. But you know, Switzerland would be a lot of fun, and I just really would like to go."

All during their conversation, Commander Colt Jackson has been holding his jacket, loosely draped over his arms. He gently unveils it and reveals a dozen, beautiful red roses.

Colt continues, "But, I'd like to ask, will you come with me?"

Lori Sanders is startled. She looks at the roses, she looks at Colt, and she throws herself into his arms. She looks up at him, surprised and says, "I didn't think you were the romantic type!"

Colt simply answers, "Remember after the meeting with Kristi Kruger we had at the NSA?"

Loris answers, her eyes glistening with tears still flowing down her cheeks, "Yes."

Colt then smiles, "I told you then that maybe one day I would surprise you!"

They come together and kiss for what seems like a lifetime and Lori is still crying. But, this time, she is crying with tears of joy and happiness.

STAY TUNED FOR MORE

COLT JACKSON

STORIES IN THE FUTURE

Book of Revelation 20:4 (King James Version):

"And I saw the souls of them that were beheaded for the witness of Jesus, and for the word of God, and which had not worshipped the beast, neither his image, neither had received his mark upon their foreheads, or in their hands; and they lived and reigned with Christ a thousand years."

Author's Personal Note

Invitation to Make a Commitment

Black Storm: Curse on the Caliphate, a work of fiction, was inspired from the author's breakout book about Commander Colt Jackson, Blood Moon Over D'Apocalypse™. The author is known for his detailed research in whatever he writes. He also scribed the non-fiction breakout work, The Federal Reserve Trilogy. It includes the The $30 Trillion Heist---Scene Of The Crime?, The $30 Trillion Heist---Follow The Money! and D'Apocalypse™ Now!---The Doomsday Cycle.

In an era when Christianity is under attack, and even splintered, all of the authors' words are targeted to reaching a broad audience with relevant, realistic stories and reporting about real-time events of our day---from a working Christian's perspective.

If you don't know Jesus Christ and you are led by the Holy Spirit, pray to Jesus for your salvation and watch the miracle of a changed life begin to take place. You will no longer fear for the future, but will find you are now in the army of God, and are one of His confident foot soldiers who KNOWS they will be victorious in the end.

If you need help in your effort to open your heart to God, the following pages might be helpful. The boxes describe three different potential situations for people. You will know, immediately, if one of them is meant for you.

God Bless You! He is mighty, indeed!!

1) FOREVER CHANGE YOUR LIFE!

"Dear God, the sole God for all eternity, I pray in your son's name, Jesus Christ, to please forgive my sins and hear my prayer. I accept Jesus Christ as my Lord and personal savior and believe you sent Him from heaven to earth, where He was born of the Virgin Mary. I believe in His life, His teachings, His horrible death and the sacrifice He made for my sins---and all mankind's sins. I KNOW AND BELIEVE HIS RESURRECTION IS TRUE AND I KNOW AND BELIEVE HE ASCENDED INTO HEAVEN AND SITS AT THE RIGHT HAND OF THE FATHER. I accept your gift and promise of eternal life through faith in Jesus Christ, for you have promised us by your grace, the grace of the one and only eternal God, I can be saved through the gift of faith. I also know my earthly works will not get me into your heavenly kingdom because you don't want any man to boast. I know we are created in Jesus to do good works out of the love and gratefulness we have for your mercy, power and gift of eternal life through the sacrifice made for us by Jesus Christ, your son and savior. AMEN"

"For it is by grace you have been saved, through faith—and this is not from yourselves, it is the gift of God—not by works, so that no one can boast. For we are God's handiwork, created in Christ Jesus to do good works, which God prepared in advance for us to do."
(Source: The Bible, Ephesians 2:8-10, New International Version).

"For God so loved the world that He gave His only begotten son, that whosever believeth in Him should not perish, but have eternal life." (Source: The Bible, John 3:16, King James Version)

If you pray this prayer, or a form of prayer accepting Jesus Christ as your Lord and Savior, I urge you to find brothers and sisters in Christ to nurture your walk with God. The Holy Spirit will guide you and you will notice a difference in the way you walk and talk through life.....you will see differences in your attitudes and actions. Embrace these, read and study His Holy Word, The Bible, and REPENT of your sins. Your heart will naturally want to do good works, because you realize what an amazing ***free*** gift has been given to you---ETERNAL LIFE!

In today's hurry-up world, it is becoming even more important to immediately embrace Jesus Christ as your savior, as God gave us Jesus to allow us to enter heaven. The Quickening is upon us all...he who has ears, let him hear....he or she who has heart, let them give it to the Lord our God—the one God of eternity. In the name of the Father, the Son and the Holy Spirit, AMEN!

Finally, after all the reading, studying, successes, failures, sacrifices and opportunities in life---there is only one thing which matters.

Jesus.

For those of you who are trapped in the sleek corporate confines of our banks and the thousands of companies thirsting for profits, worldwide, this book, is especially for you. It is written to those living in comfort in the most American and Western-European sense of the word, with children and young adults honey-stuck to their smart phones and video games, while mom and dad are living, eating and drinking like kings and queens, with everyone engrossed in the TV set. Unfortunately for one all, they are completely ignoring the dark, ominous clouds on the horizon foreshadowing their very own deaths. Without Jesus you are doomed.

To any and all backsliders who are out there—I urge you to read the Holy Scripture and take heed and warning. Take up the courage of conviction and KNOW The Bible is TRUE!! The evidence is overwhelming. Pray for the strength to overcome evil with good AND DO SO!!

2) **THE BACKSLIDER'S MAKE OR BREAK TIME!**

"What good is it, my brothers and sisters, if someone claims to have faith but has no deeds? Can such faith save them? Suppose a brother or a sister is without clothes and daily food. If one of you says to them, "Go in peace; keep warm and well fed," but does nothing about their physical needs, what good is it? In the same way, faith by itself, if it is not accompanied by action, is dead.

But someone will say, "You have faith; I have deeds."

Show me your faith without deeds, and I will show you my faith by my deeds. You believe that there is one God. Good! Even the demons believe that—and shudder.

You foolish person, do you want evidence that faith without deeds is useless? Was not our father Abraham considered righteous for what he did when he offered his son Isaac on the altar? You see that his faith and his actions were working together, and his faith was made complete by what he did. And the scripture was fulfilled that says, "Abraham believed God, and it was credited to him as righteousness," and he was called God's friend. You see that a person is considered righteous by what they do and not by faith alone.

In the same way, was not even Rahab the prostitute considered righteous for what she did when she gave lodging to the spies and sent them off in a different direction? As the body without the spirit is dead, so faith without deeds is dead."
(Source: The Bible, James 2:14-24, New International Version).

Sadly, there are millions of souls who believe in Jesus, but go back to their old ways of sin after they have received Jesus as their savior and don't do any good works. The Holy Scripture is very clear on this topic. FAITH WITHOUT WORKS IS DEAD.

As even the History Channel recounts the story of Abraham being obedient to God, The Bible—in both the Old and New Testaments---informed us of God's principle of obedience long before it streamed into our living rooms courtesy of your cable provider. If you truly have given your heart and life to Jesus Christ, your natural desire---and your own predetermined destiny, is to do good deeds!

Backsliders, one and all, you must believe in the imminence of His arrival and I implore you to make a different, real change in your life.

Remember…***"Even the demons believe...and they shudder."***

It is not enough to just "believe." Make a difference in the life around you---just as Jesus and the disciples did, and "do" for other people. Help them, pray for them, feed them, cloth them, return to a simpler life (e.g., Jesus and His disciples didn't have any worldly possessions). Do not go down the slippery slope of, "Well, the Lord blessed me with this, so it is all OK…."

We probably all know numbers of wealthy Christians who have been blessed with much---but have done very little for other people in relation to what was given to them. To these people, I implore you to remember what Jesus said:

> **"It is easier for a camel to go through the eye of a needle, than for a rich man to enter into the kingdom of God."**
> (Source: The Bible, Matthew 19:23-24, King James Version).

Of course, the wealthy will go on to the next words of Jesus when the disciples asked Him who can be saved and Jesus replied all things are possible for God.

The admonition is clear, however, the wealthy shall be scrutinized very carefully by our heavenly and just God.

The disciples themselves gave up EVERYTHING THEY HAD TO FOLLOW JESUS. They all changed. They were all afraid at His death and went into hiding (fearful the Romans and the Jews would crucify them, just as they had Jesus), but they became ferocious and vociferous zealots for Christ---brandishing the helmet of salvation, the shield of faith, while swinging with all of their might, the sword of the Holy Spirit. They died martyrs' deaths, with one spared to write the Book of Revelation---the Book of the Apocalypse, who of course was the Apostle John.

It was quite a change for a small group of men---all once terrified and in hiding, yet every one of them became unafraid to die for Christ. I don't know of a single case in history where any person, let alone an entire group of people (except John, spared to write the Book of Revelation, the Book of the Apocalypse) would die for a lie when they could have just walked away, if they had just chosen to denounce Jesus Christ as their Lord and Savior!

These men are as relevant to us today, more than ever before in history.

If you have ears to hear, then listen; if you have eyes to see, then read:

3) THE MOST DANGEROUS ROUTE TO HELL IS BEING TOO COMFORTABLE!

> **"Then Peter said to Him, 'Behold, we have left everything and followed You; what then will there be for us?' And Jesus said to them, 'Truly I say to you, that you who have followed Me, in the regeneration when the Son of Man will sit on His glorious throne, you also shall sit upon twelve thrones, judging the twelve tribes of Israel. And everyone who has left houses or brothers or sisters or father or mother or children or farms for My name's sake, will receive many times as much, and will inherit eternal life. But many who are first will be last; and the last, first.'"**
> (Source: The Bible, Matthew 19:27-30, New American Standard Version).

The author, Robert Lyman Peter Kelly (as a boy, I chose "Peter" as my confirmation name), just like you, the reader, and just like the Apostle Peter, is a sinner and the words I write are only possible because I have been forgiven by the grace of God through the faith I have in the sacrifice and resurrection of the son of God, Jesus Christ. I know He shed His blood for me and for all mankind. He was born of the Virgin Mary and Jesus truly lived, taught, breathed, witnessed, suffered, died, buried, was resurrected AND ascended into heaven in front of many, many witnesses--all in fulfillment of the scriptures. He performed (and performs today) many miracles for the living while He walked among the disciples—and continues to perform those miracles after His earthly death.

The biggest miracle Jesus gives us ALL is the miraculous change He brings to those who truly give themselves to Him. This is the miracle of a changed life. If you are not saved through the blood of Christ and your absolute faith in His resurrection and teachings, pray to God as you have been led by the Holy Spirit in one of the DECISION BOXES of this book, above, or otherwise. Ask for His forgiveness in His son's name, Jesus Christ, and commit your life to Him. SEEK other like-minded individuals who can help you learn more about your new walk and birth in Jesus. This is what

Jesus meant when He said you cannot enter the Kingdom of Heaven---UNLESS YOU ARE BORN AGAIN!!

> Finally, for all you backsliders out there, whether RICH or POOR, **Jesus did not put a limit on how many times you can ask for, AND RECEIVE, complete forgiveness and thus eternal life in Heaven.**

With your new life, you have a new beginning. Make the most of it---for His glory and not your own. When the disaster of D'Apocalypse™ strikes, remain calm and KNOW the Lord is our God and has complete control of EVERYTHING. This is part of the battle which must be fought before Jesus returns to earth.

I hope and pray you choose the side of might for right. May that prayer and focus on the good Lord, above, be on all of the men and women in uniform, particularly, who protect our nation. We must return to a Godly focus and beg for His forgiveness.

To this, the people said, in Jesus name,

Amen!

Christian Cross in the Roman Coliseum

Made in the USA
Monee, IL
13 September 2026

62979537R00193